Sweet Dreams

The Jean Nicod Lectures
François Recanati, editor

Sweet Dreams

Philosophical Obstacles to a Science of Consciousness

Daniel C. Dennett

A Bradford Book
The MIT Press
Cambridge, Massachusetts
London, England

First MIT Press paperback edition, 2006

MIT Press books may be purchased at special quantity discounts for business or sales promotional use. For information, please email special_sales@mitpress.mit.edu or write to Special Sales Department, The MIT Press, 55 Hayward Street, Cambridge, MA 02142.

This book was set in Stone Sans and Stone Serif by SNP Best-set Typesetter Ltd., Hong Kong, and was printed and bound in the United States of America.

Library of Congress Cataloging-in-Publication Data

Dennett, Daniel Clement.
Sweet dreams : philosophical obstacles to a science of consciousness / Daniel C. Dennett.
    p.   cm.—(The Jean Nicod lectures)
"A Bradford book."
Includes bibliographical references and index.
ISBN-10: 0-262-04225-8 (hc : alk. paper)—0-262-54191-2 (pb : alk. paper)
ISBN-13: 978-0-262-04225-3 (hc : alk. paper)—978-0-262-54191-6 (pb : alk. paper)
1. Consciousness. I. Title. II. Series.

B945.D393S94   2005
153—dc22

                                                        2004048681

10   9   8   7   6   5

Contents

Series Foreword

The Jean Nicod Lectures are delivered annually in Paris by a leading philosopher of mind or philosophically oriented cognitive scientist. The 1993 inaugural lectures marked the centenary of the birth of the French philosopher and logician Jean Nicod (1893–1931). The lectures are sponsored by the Centre National de la Recherche Scientifique (CNRS) and are organized in cooperation with the Fondation Maison des Sciences de l'Homme (MSH Foundation). The series hosts the texts of the lectures or the monographs they inspire.

Jacques Bouveresse, President of the Jean Nicod Committee
François Recanati, Secretary of the Jean Nicod Committee and Editor of the Series

Jean Nicod Committee
Mario Borillo
Jean-Pierre Changeux
Jean-Gabriel Ganascia

André Holley
Michel Imbert
Pierre Jacob
Jacques Mehler
Elisabeth Pacherie
Philippe de Rouilhan
Dan Sperber

Preface

Several years ago, I was invited to write a review surveying the recent theoretical work on consciousness by authors in several fields, ranging from quantum physics and chemistry through neuroscience and psychology to philosophy and literature. A swift survey of the *good* new books lined up on top of my book-shelf (it numbers 78, today, January 28, 2004, and those are just the books) persuaded me that I should beg off the job. It has been a tumultuous decade, so rambunctious that several people are writing books just about the tumult. I am adding this book to the flood as an exercise in deferred maintenance. The theory I sketched in *Consciousness Explained* in 1991 is holding up pretty well, I think, in spite of major advances in empirical outlook (on the positive side) and several waves of misconstrual (on the negative). I didn't get it all right the first time, but I didn't get it all wrong either. It is time for some revision and renewal.

I spent much of the early 1990s responding to the criticisms and other reactions[1] my book had provoked before turning my

1. A partial list can be found at http://sun3.lib.uci.edu/~scctr/philosophy/dennett/.

attention to the neo-Darwinian theory of evolution and its philosophical implications. After the publication of *Darwin's Dangerous Idea* in 1995, I spent several more years defending and expanding *its* claims, while the literature on consciousness burgeoned apace. As the century turned, I knew I had to go back to the issues raised by my 1991 book and refine my positions in response to new waves of empirical results and theoretical proposals and challenges. A series of essays resulted. The first chapter of this book was my Millennial Lecture to the Royal Institute of Philosophy, giving my opinion on the state of play in the philosophy of mind at the turn of the century. It was subsequently published (Dennett 2001b). In November, 2001, I gave the Jean Nicod Lectures at the Institut Nicod in Paris, on philosophical obstacles to a science of consciousness, and the following November I presented a revised and expanded version of those lectures as the Daewoo Lectures in Seoul. Chapters two through five of this book are drawn, with further revisions, from those presentations. (A version of one of the Nicod Lectures was incorporated into chapter 8 of Dennett 2003a and does not appear here, and a version of chapter five of this volume is also being published in Alter 2005.) Chapter 6 is reprinted from *Cognition* (Dennett 2001a), chapter 7 draws on a lecture of mine in London in 1999, and chapter 8 is a short essay on consciousness forthcoming in Richard Gregory's revised edition of the *Oxford Companion to the Mind*. There are a few stylistic revisions in the chapters published or forthcoming elsewhere.

The Multiple Drafts Model of consciousness is also a model of my academic life during the last dozen years. Giving several dozen public lectures a year on consciousness to widely different audiences encourages a large amount of adaptation and mutation of previously used material. In this volume I have attempted to freeze time somewhat arbitrarily and compose a "best" version of all this, trying to minimize repetition while preserving context. That is just what we do, according to my theory, when we tell others—or even our later selves—about our conscious experience. Further postmillennial essays of mine on consciousness were not included because they either contain earlier versions of the discussions in the present chapters or are responses to specific essays or books and really need to be read in their original context:

"Explaining the 'Magic' of Consciousness," in *Exploring Consciousness, Humanities, Natural Science, Religion*, Proceedings of the International Symposium, Milano, November 19–20, 2001 (published in December, 2002, Fondazione Carlo Erba), pp. 47–58; reprinted in J. Laszlo, T. Bereczkei, C. Pleh, eds., *Journal of Cultural and Evolutionary Psychology*, 1, 2003, pp. 7–19 (Dennett 2001c).

"Who's on First? Heterophenomenology Explained," *Journal of Consciousness Studies*, special issue: Trusting the Subject? (Part 1), 10, no. 9–10, October 2003, pp. 19–30; also appears in A. Jack and A. Roepstorff eds., *Trusting the Subject?* volume 1, Imprint Academic, 2003, pp. 19–30 (Dennett 2003b).

"The Case for Rorts," in *Rorty and His Critics*, R. B. Brandom, ed., Blackwell, 2000, pp. 89–108 (Dennett 2000a).

"It's Not a Bug, It's a Feature," commentary on Humphrey, *Journal of Consciousness Studies*, 7, 2000, pp. 25–27 (Dennett 2000b).

"Surprise, Surprise," commentary on O'Regan and Noë, *Behavioral and Brain Sciences*, 24:5, 2001, p. 982 (Dennett 2001d).

"How Could I Be Wrong? How Wrong Could I Be?" for special issue of *Journal of Consciousness Studies*, "Is The Visual World a Grand Illusion?", ed. Alva Noë, vol. 9, no. 5–6, January 13, 2002, pp. 13–16 (Dennett 2002a).

"Does Your Brain Use the Images in It, and If So, How?" commentary on Pylyshyn, *Behavioral and Brain Sciences*, 25:2, 2002, pp. 189–190 (Dennett 2002b).

"Look Out for the Dirty Baby," commentary on Baars, *Journal of Consciousness Studies*, "The Double Life of B. F. Skinner," 10:1, 2003, pp. 31–33 (Dennett 2003c).

"Making Ourselves at Home in Our Machines," review of Wegner, *The Illusion of Conscious Will*, The MIT Press, 2002, in *Journal of Mathematical Psychology* 47, 2003, pp. 101–104 (Dennett 2003d).

Readers hoping to find me taking sides in various ongoing scientific controversies in the cognitive science of consciousness will be disappointed, for while I have strong opinions about many of these issues, I am mostly resisting the temptation to go out on all my favorite empirical limbs here, since I want to focus attention on the *philosophical* issues that continue to bedevil the field, confusing and distracting philosophers and nonphilosophers alike. I have always thought that John Locke led the way with his relatively modest vision of a philosopher's proper role, in the "Epistle to the Reader" at the beginning of his *Essay Concerning Human Understanding*, 1690. We have not yet reached consensus on who, if anybody, is the Isaac Newton or even the Christian Huygens of cognitive science, but aside from that, Locke's words leap three centuries with uncanny accuracy:

. . . in an age that produces such masters as the great Huygenius and the incomparable Mr. Newton, with some others of that strain, it is ambition enough to be employed as an under-labourer in clearing the ground a little, and removing some of the rubbish that lies in the way to knowledge;—which certainly had been very much more advanced in the world, if the endeavours of ingenious and industrious men had not been much cumbered with the learned but frivolous use of uncouth, affected, or unintelligible terms, introduced into the sciences, and there made an art of, to that degree that Philosophy, which is nothing but the true knowledge of things, was thought unfit or incapable to be brought into well-bred company and polite conversation.

I am grateful to many colleagues, students, critics, reviewers, audiences, and correspondents for their insights and provocations. In particular, I am grateful to Tufts University for supporting the home for my work, the Center for Cognitive Studies, and to the Institut Jean Nicod and the Daewoo Foundation for sponsoring the lectures that form the heart of this book. I want to thank Al Seckel for helping me find the best visual illustration for the book (see fig. 2.2). And as always, I want to thank my wife of more than forty years for the advice, support, companionship and love without which I couldn't manage at all.

Daniel C. Dennett
January 29, 2004

1 The Zombic Hunch: Extinction of an Intuition?

The Steinberg cartoon on the cover shows one good way of looking at the problem of consciousness. If this is the metaphorical truth about consciousness, what is the literal truth? What is going on in the world (largely in this chap's brain, presumably) that makes it the case that this gorgeous metaphor is so apt?

1 The Naturalistic Turn

Our conception of this question at the end of the twentieth century is strikingly different from the ways we might have thought about the same issue at the beginning of the century, thanks very little to progress in philosophy and very much to progress in science. Steinberg's *pointillist* rendering of our conscious man gives us a fine hint about the major advances in outlook that promise—to many of us—to make all the difference. What we now know is that each of us is an assemblage of trillions of cells, of thousands of different sorts. Most of the cells that compose your body are descendants of the egg and sperm cell whose union started you (there are also millions of hitchhikers from thousands of different lineages stowed away in your body), and, to put it vividly and bluntly, *not a single one of the cells that compose you knows who you are, or cares.*

The individual cells that compose you are alive, but we now understand life well enough to appreciate that each cell is a mindless mechanism, a largely autonomous microrobot, no more conscious than a yeast cell. The bread dough rising in a bowl in the kitchen is teeming with life, but nothing in the bowl is sentient or aware—or if it is, then this is a remarkable fact for which, at this time, we have not the slightest evidence. For we now know that the "miracles" of life—metabolism, growth, self-repair, self-defense, and, of course, reproduction—are all accomplished by dazzlingly intricate, but nonmiraculous, means. No sentient supervisor is needed to keep metabolism going, no *élan vital* is needed to trigger self-repair, and the incessant nano-factories of replication churn out their duplicates without any help from ghostly yearnings or special life forces. A hundred kilos of yeast does not wonder about Braque, or about anything, but you do, and you are made of parts[1] that are fundamentally the same sort of thing as those yeast cells, only with different tasks to perform. Your trillion-robot team is gathered together in a breathtakingly efficient regime that has no dictator but manages to keep itself organized to repel outsiders, banish the weak, enforce iron rules of discipline—and serve as the head-quarters of one conscious self, one mind. These communities of cells are fascistic in the extreme, but *your* interests and values have almost nothing to do with the limited goals of the cells that compose you—fortunately. Some people are gentle and generous, others are ruthless; some are pornographers and others devote their lives to the service of God; and it has been

1. Eukaryotic cells.

tempting over the ages to imagine that these striking differences must be due to the special features of some *extra* thing—a soul—installed somehow in the bodily headquarters. Until fairly recently, this idea of a rather magical extra ingredient was the only candidate for an explanation of consciousness that even *seemed* to make sense. For many people, this idea (dualism) is *still* the only vision of consciousness that makes any sense to them, but there is now widespread agreement among scientists and philosophers that dualism is—must be—simply false: we are each *made of* mindless robots and nothing else, no non-physical, nonrobotic ingredients at all.

But how could this possibly be? More than a quarter of a millennium ago, Leibniz posed the challenge to our imaginations with a vivid intuition pump, a monumentally misleading grandfather to all the Chinese Rooms (Searle), Chinese Nations (Block) and latter-day zombies.

> Moreover, it must be confessed that *perception* and that which depends upon it are *inexplicable on mechanical grounds*, that is to say, by means of figures and motions. And supposing there were a machine, so constructed as to think, feel, and have perception, it might be conceived as increased in size, while keeping the same proportions, so that one might go into it as into a mill. That being so, we should, on examining its interior, find only parts which work one upon another, and never anything by which to explain a perception. Thus it is in a simple substance, and not in a compound or in a machine, that perception must be sought for. (Leibniz, *Monadology*, 1714: para. 17 [Latta translation])

There is a striking non sequitur in this famous passage, which finds many echoes in today's controversies. Is Leibniz's claim epistemological—we'll *never understand* the machinery of consciousness—or metaphysical—consciousness *couldn't be* a matter of "machinery"? His preamble and conclusion make it plain that he took himself to be demonstrating a metaphysical truth, but the only grounds he offers would—at best—support the more modest epistemological reading.[2] Somebody *might* have used Leibniz's wonderful Gulliverian image to *illustrate and render plausible*[3] the claim that although consciousness is—must be, in the end—a product of some gigantically complex mechanical system, it will surely be utterly beyond anybody's intellectual powers to explain how this is so. But Leibniz clearly intends us to treat his example as demonstrating the absurdity of the very idea that consciousness could be such an emergent effect of a hugely complex machine ("Thus it is in a simple substance, and not in a compound or in a machine, that perception must be sought for").

The same mismatch between means and ends haunts us today: Noam Chomsky, Thomas Nagel, and Colin McGinn

---

2. Leibniz makes this particularly clear in another passage quoted in Latta's translation:

> If in that which is organic there is nothing but mechanism, that is, bare matter, having differences of place, magnitude and figure; nothing can be deduced or explained from it, except mechanism, that is, except such differences as I have mentioned. For from anything taken by itself nothing can be deduced and explained, except differences of the attributes which constitute it. Hence we may readily conclude that in no mill or clock as such is there to be found any principle which perceives what takes place in it; and it matters not whether the things contained in the "machine" are solid or fluid or made up of both. Further we know that there is no essential difference between coarse and fine bodies, but only a difference of magnitude. Whence it follows that, if it is inconceivable how perception arises in any coarse "machine," whether it be made up of fluids or solids, it is equally inconceivable how perception can arise from a fine "machine"; for if our senses were finer, it would be the same as if we were perceiving a coarse "machine," as we do at present. (From *Commentatio de Anima Brutorum*, 1710, quoted in Latta, p. 228)

3. It would not, of course, *prove* anything at all. It is just an intuition pump.

(among others) have all surmised, or speculated, or claimed, that consciousness is beyond all human understanding, a mystery not a puzzle, to use Chomsky's proposed distinction.[4] According to this line of thought, we lack the wherewithal—the brain power, the perspective, the intelligence—to grasp *how* the "parts which work one upon another" could constitute consciousness. Like Leibniz, however, these thinkers have also hinted that they themselves understand the mystery of consciousness a little bit—just well enough to able to conclude that it *couldn't* be solved by any mechanistic account. And, just like Leibniz, they have offered nothing, really, in the way of arguments for their pessimistic conclusions, beyond a compelling image. When they contemplate the prospect they simply draw a blank, and thereupon decide that no further enlightenment lies down that path *or could possibly* lie down that path.

Might it be, however, that Leibniz, lost in his giant mill, just couldn't see the woods for the trees? Might there not be a bird's-eye view—*not* the first-person perspective of the subject in question, but a higher-level *third*-person perspective—from which, if one squinted just right, one could bring into focus the recognizable patterns of consciousness in action? Might it be that somehow the *organization* of all the parts which work one upon another yields consciousness as an emergent product? And if so, why couldn't we hope to understand it, once we had developed the right concepts? This is the avenue that has been enthusiastically and fruitfully explored during the last quarter century under the twin banners of cognitive science and functionalism—the extrapolation of *mechanistic naturalism* from the

4. Most recently, in the following works: Chomsky 1994; Nagel 1998; McGinn 1999.

body to the mind. After all, we have now achieved excellent mechanistic explanations of metabolism, growth, self-repair, and reproduction, which not so long ago also looked too marvelous for words. Consciousness, on this optimistic view, is indeed a wonderful thing, but not *that* wonderful—not too wonderful to be explained using the same concepts and perspectives that have worked elsewhere in biology.

Consciousness, from this perspective, is a relatively recent fruit of the evolutionary algorithms that have given the planet such phenomena as immune systems, flight, and sight. In the first half of the century, many scientists and philosophers might have agreed with Leibniz about the mind, simply because the mind seemed to consist of phenomena *utterly unlike* the phenomena in the rest of biology. The inner lives of mindless plants and simple organisms (and our bodies below the neck) might yield without residue to normal biological science, but nothing remotely mindlike could be accounted for in such mechanical terms. Or so it must have seemed until something came along in midcentury to break the spell of Leibniz's intuition pump. Computers. Computers are mindlike in ways that no earlier artifacts were: they can control processes that perform tasks that call for discrimination, inference, memory, judgment, anticipation; they are generators of new knowledge, finders of patterns—in poetry, astronomy, and mathematics, for instance—that heretofore only human beings could even hope to find. We now have real-world artifacts that dwarf Leibniz's giant mill both in speed and intricacy. And we have come to appreciate that what is well nigh invisible at the level of the meshing of billions of gears may nevertheless be readily comprehensible at higher levels of analysis—at any of many nested "software" levels, where the patterns of patterns of patterns of organization (of organization of organization) can render salient *and explain* the marvelous competences of the mill. The sheer existence of computers has provided an existence proof of undeniable influence: *there are* mechanisms—brute, unmysterious mechanisms operating according to routinely well-understood physical principles—that have many of the competences heretofore assigned only to minds.

One thing we know to a moral certainty about computers is that there is nothing up their sleeves: no ESP or morphic resonance between disk drives, no action-at-a-distance accomplished via strange new forces. The *explanations* of whatever talents computers exhibit are models of transparency, which is one of the most attractive features of cognitive science: we can be quite sure that *if* a computational model of *any* mental phenomenon is achieved, it will inherit this transparency of explanation from its simpler ancestors.

In addition to the computers themselves, wonderful exemplars and research tools that they are, we have the wealth of new concepts computer science has defined and made familiar. We have learned how to think fluently and reliably about the cumulative effects of intricate cascades of micromechanisms, trillions upon trillions of events of billions of types, interacting on dozens of levels. Can we harness these new powers of disciplined imagination to the task of climbing out of Leibniz's mill? The hope that we can is, for many of us, compelling—even inspiring. We are quite certain that a naturalistic, mechanistic explanation of consciousness is not just possible; it is fast becoming actual. It will just take a lot of hard work of the sort that has been going on in biology all century, and in cognitive science for the last half century.

2 The Reactionaries

But in the last decade of the century a loose federation of reactionaries has sprung up among philosophers in opposition to this evolutionary, mechanistic naturalism. As already noted, there are the *mysterians*, Owen Flanagan's useful term for those who not only find this optimism ill founded but also think that defeat is certain. Then there are those who are not sure the problem is insoluble, but do think that they can titrate the subtasks into the "easy problems" and the "Hard Problem" (David Chalmers) or who find what they declare to be an "Explanatory Gap" (Joseph Levine) that has so far—and perhaps always will—defy those who would engulf the mind in one unifying explanation.[5] A curious anachronism found in many but not all of these reactionaries is that to the extent that they hold out any hope at all of solution to the problem (or problems) of consciousness, they speculate that it will come not from biology or cognitive science, but from—of all things—physics!

One of the first to take up this courtship with physics was David Chalmers, who suggested that a theory of consciousness should "take experience itself as a fundamental feature of the world, alongside mass, charge, and space-time." As he correctly noted, "No attempt is made [by physicists] to explain these features in terms of anything simpler,"[6] a theme echoed by Thomas Nagel:

> Consciousness should be recognized as a conceptually irreducible aspect of reality that is necessarily connected with other equally

5. Chalmers 1995, 1996; Levine 1983.
6. Chalmers 1995.

irreducible aspects—as electromagnetic fields are irreducible to but necessarily connected with the behaviour of charged particles and gravitational fields with the behaviour of masses, and vice versa.[7]

And by Noam Chomsky:

The natural conclusion . . . is that human thought and action are properties of organized matter, like "powers of attraction and repulsion," electrical charge, and so on.[8]

And by Galen Strawson, who says, in a review of Colin McGinn's most recent book: "we find consciousness mysterious only because we have a bad picture of matter" and adds:

We have a lot of mathematical equations describing the behavior of matter, but we don't really know anything more about its intrinsic nature. The only other clue that we have about its intrinsic nature, in fact, is that when you arrange it in the way that it is arranged in things like brains, you get consciousness.[9]

Not just philosophers and linguists have found this an attractive idea. Many physicists have themselves jumped on the bandwagon, following the lead of Roger Penrose, whose speculations about quantum fluctuations in the microtubules of neurons have attracted considerable attention and enthusiasm in spite of a host of problems.[10] What all these views have in common

7. Nagel, "Conceiving the Impossible," p. 338.

8. Chomsky, "Naturalism and Dualism," p. 189. Chomsky is talking about the conclusion drawn by La Mettrie and Priestley, but his subsequent discussion, footnoting Roger Penrose and John Archibald Wheeler, makes it clear that he thinks this is a natural conclusion today, not just in early post-Newtonian days.

9. Strawson 1999.

10. Incurable optimist that I am, I find this recent invasion by physicists into the domains of cognitive neuroscience to be a cloud with a silver lining: for the first time in my professional life, an interloping

is the idea that some revolutionary principle of physics could be a *rival* to the idea that consciousness is going to be explained in terms of "parts which work one upon another," as in Leibniz's mill.

Suppose they are right. Suppose the Hard Problem—whatever it is—can be solved only by confirming some marvelous new and irreducible property of the *physics* of the cells that make up a brain. One problem with this is that the physics of your brain cells is, so far as we know, the same as the physics of those yeast cells undergoing population explosion in the dish. The differences in functionality between neurons and yeast cells are explained in terms of differences of cell anatomy or cytoarchitecture, not physics. Could it be, perhaps, that those differences in anatomy permit neurons to respond to physical differences to which yeast cells are oblivious? Here we must tread carefully, for if we don't watch out, we will simply reintroduce Leibniz's baffling mill at a more microscopic level—watching the quantum fluctuations in the microtubules of a single cell and not being able to see how any amount of *those* "parts which work one upon another" could explain consciousness.

If you want to avoid the bafflement of Leibniz's mill, the idea had better be, instead, that consciousness is an irreducible

---

discipline beats out philosophy for the prize for combining arrogance with ignorance about the field being invaded. Neuroscientists and psychologists who used to stare glassy-eyed and uncomprehending at philosophers arguing about the fine points of *supervenience* and *intensionality-with-an-s* now have to contend in a similar spirit with the arcana of *quantum entanglement* and *Bose–Einstein condensates*. It is tempting to suppose that as it has become harder and harder to make progress in physics, some physicists have sought greener pastures where they can speculate with even less fear of experimental recalcitrance or clear contradiction.

property that inheres, somehow "in a simple substance," as Leibniz put it, "and not in a compound or in a machine." So let us suppose that, thanks to their physics, neurons enjoy a tiny smidgen (a quantum, perhaps!) of consciousness. We will then have solved the problem of how large ensembles of such cells— such as you and I—are conscious: we are conscious because our brains are made of the right sort of stuff, stuff with the micro-*je-ne-sais-quoi* that is needed for consciousness. But even if we had solved *that* problem, we would still have the problem illustrated by my opening illustration: how can cells, even *conscious* cells, that themselves know nothing about art or dogs or mountains compose themselves into a thing that has conscious thoughts about Braque or poodles or Kilimanjaro? How can the whole ensemble be so knowledgeable of the passing show, so in touch with distal art objects (to say nothing of absent artists and mountains) when all of its parts, however conscious or sentient they are, are myopic and solipsistic in the extreme? We might call this the *topic*-of-consciousness question.[11]

I suspect that this turn to physics looks attractive to some people mainly because they have not yet confronted the need to answer *this* question, for once they do attempt it, they find that a "theory" that postulates some fundamental and irreducible sentience-field or the like has no resources *at all* to deal with it. *Only* a theory that proceeds in terms of how the parts

11. A classic example of the topic problem in nature, and its ultimately computational solution, is Douglas Hofstadter's famous "Prelude . . . Ant Fugue" in *Gödel, Escher, Bach* (1979), the dialogue comparing an ant colony ("Aunt Hillary") to a brain, whose parts are equally clueless contributors to systemic knowledge of the whole. In his reflections following the reprinting of this essay in Hofstadter and Dennett, eds., *The Mind's I* (1981), he asks "Is the soul more than the hum of its parts?"

work together in larger ensembles has any hope of shedding light on the topic question, and once theory has ascended to such a high level, it is not at all clear what use the lower-level physical sophistications would be. Moreover, there already are many models of systems that uncontroversially answer *versions* of the topic question, and they are all computational. How can the little box on your desk, whose parts know nothing at all about chess, beat you at chess with such stunning reliability? How can the little box driving the pistons attached to the rudder do a better job of steering a straight course than any old salt with decades at sea behind him? Leibniz would have been ravished with admiration by these mechanisms, which would have shaken his confidence—I daresay—in the claim that no mechanistic explanation of "perception" was possible.

David Chalmers, identifier of the Hard Problem, would agree with me, I think. He would classify the topic question as one of the "easy problems"—one of the problems that *does* find its solution in terms of computational models of control mechanisms. It follows from what he calls the principle of organizational invariance.[12] Consider once again our *pointillist* gentleman and ask if we can tell from the picture whether he's a genuinely conscious being or a zombie—a philosopher's zombie that is behaviorally indistinguishable from a normal human being but is utterly lacking in consciousness. Even the zombie version of this chap would have a head full of dynamically interacting data-structures, with links of association bringing their sequels online, suggesting new calls to memory, composing on the fly new structures with new meanings and powers. Why? Because only a being with such a system of internal operations and

12. Chalmers 1996, esp. chapter 7.

activities could nonmiraculously maintain the complex set of behaviors this man would no doubt exhibit if we put him to various tests. If you want a theory of all that information-processing activity, it will have to be a computational theory, whether or not the man is conscious. According to Chalmers, where normal people have a stream of consciousness, zombies have a stream of unconsciousness, and he has argued persuasively that whatever explained the *purely informational competence* of one (which includes every transition, every construction, every association depicted in this thought balloon) would explain the same competence in the other. Since the literal truth about the mechanisms responsible for all the sworls and eddies in the stream, as well as the informational contents of the items passing by, is—ex hypothesi—utterly unaffected by whether or not the stream is conscious or unconscious, Steinberg's cartoon, a brilliant metaphorical rendering of consciousness, is exactly as good a metaphorical rendering of what is going on inside a zombie. (See, e.g., the discussion of zombie beliefs in Chalmers 1996, pp. 203–205.)

3   An Embarrassment of Zombies

Must we talk about zombies? Apparently we must. There is a powerful and ubiquitous intuition that computational, mechanistic models of consciousness, of the sort we naturalists favor, *must leave something out*—something important. Just what must they leave out? The critics have found that it's hard to say, exactly: qualia, feelings, emotions, the what-it's-likeness (Nagel)[13] or the ontological subjectivity (Searle)[14] of conscious-

13. Nagel 1974.
14. Searle 1992.

ness. Each of these attempts to characterize the phantom residue has met with serious objections and been abandoned by many who nevertheless want to cling to the intuition, so there has been a gradual process of distillation, leaving just about all the reactionaries, for all their disagreements among themselves, united in the conviction *that there is a real difference between a conscious person and a perfect zombie*—let's call that intuition the *Zombic Hunch*—leading them to the thesis of *Zombism*: that *the fundamental flaw in any mechanistic theory of consciousness is that it cannot account for this important difference.*[15]

A hundred years from now, I expect this claim will be scarcely credible, but let the record show that in 1999, John Searle, David Chalmers, Colin McGinn, Joseph Levine and many other philosophers of mind don't just *feel the tug* of the Zombic Hunch (I can feel the tug as well as anybody), they *credit* it. They are, however reluctantly, Zombists, who maintain that the zombie challenge is a serious criticism. It is not that they don't recognize the awkwardness of their position. The threadbare stereotype of philosophers passionately arguing about how many angels can dance on the head of a pin is not much improved when the topic is updated to whether zombies—admitted by all to be imaginary beings—are (1) metaphysically impossible, (2) logically impossible, (3) physically impossible, or just (4) extremely unlikely to exist. The reactionaries have acknowledged that many who take zombies seriously have simply failed to imagine the prospect correctly. For instance, if you were *surprised* by my claim that the Steinberg cartoon would be an

15. In the words of one of their most vehement spokespersons, "It all comes down to zombies" (Selmer Bringsjord, "Dennett versus Searle on Cognitive Science: It All Comes Down to Zombies and Searle Is Right" (paper presented at the APA, December, 1994).

equally apt metaphorical depiction of the goings on in a zombie's head, you had not heretofore understood what a zombie is (and isn't). More pointedly, if you *still* think that Chalmers and I are just wrong about this, you are simply operating with a mistaken concept of zombies, one that is irrelevant to the philosophical discussion. (I mention this because I have found that many onlookers, scientists in particular, have a hard time believing that philosophers can be taking such a preposterous idea as zombies seriously, so they generously replace it with some idea that one *can* take seriously—but one that does not do the requisite philosophical work. Just remember: by definition, a zombie behaves *indistinguishably* from a conscious being—in all possible tests, including not only answers to questions [as in the Turing test] but psychophysical tests, neurophysiological tests—all tests that any "third-person" science can devise.)

Thomas Nagel is one reactionary who has recoiled somewhat from zombies. In his recent address to this body, Nagel is particularly circumspect in his embrace. On the one hand, he declares that naturalism has so far failed us:

> We do not at present possess the conceptual equipment to understand how subjective and physical features could both be essential aspects of a single entity or process.

Why not? Because "we still have to deal with the apparent conceivability of . . . a zombie." Notice that Nagel speaks of the *apparent* conceivability of a zombie. I have long claimed that this conceivability is *only* apparent; some misguided philosophers *think* they can conceive of a zombie, but they are badly mistaken.[16] Nagel, for one, agrees:

16. Dennett 1991, esp. chapters 10–12; 1994a; 1995b.

the powerful intuition that it is conceivable that an intact and normally functioning physical human organism could be a completely unconscious zombie is an illusion.[17]

David Chalmers is another who is particularly acute in his criticisms of the standard mis-imaginations that are often thought to support the zombie challenge (his 1996, chapter 7, "Absent Qualia, Fading Qualia, Dancing Qualia," bristles with arguments against various forlorn attempts), but in the end, he declares that although zombies are in every realistic sense impossible, we "nonreductive functionalists" still leave something out—or rather, we leave a job undone. We cannot provide "*fundamental* laws" from which one can deduce that zombies are impossible (p. 276 and elsewhere). Chalmers's demand for fundamental laws lacks the independence he needs if he is to support his crediting of the Zombic Hunch, for it *arises from* that very intuition: *if* you believe that consciousness sunders the universe in twain, into those things that have it and those that don't, *and* you believe this is a fundamental metaphysical distinction, then the demand for fundamental laws that enforce and explain the sundering makes some sense, but we naturalists think that this elevation of consciousness is itself suspect, supported by tradition and nothing else. Note that nobody these days would clamor for fundamental laws of *the theory of kangaroos*, showing why pseudo-kangaroos are physically, logically, metaphysically impossible. Kangaroos are wonderful, but not *that* wonderful. We naturalists think that consciousness, like locomotion or predation, is something that comes in different varieties, with some shared functional properties, but many differences, owing to different evolutionary histories and

17. Nagel, "Conceiving the Impossible," p. 342.

circumstances. We have no use for fundamental laws in making these distinctions.

We are all *susceptible* to the Zombic Hunch, but if we are to credit it, we need a good argument, since the case has been made that it is a persistent cognitive illusion and nothing more. I have found no good arguments, and plenty of bad ones. So why, then, do so many philosophers persist in their allegiance to an intuition that they themselves have come to see is of suspect provenance? Partly, I think, this is the effect of some serious misdirection that has bedeviled communication in cognitive science in recent years.

4 Broad Functionalism and Minimalism

Functionalism is the idea that handsome is as handsome does, that matter matters only because of what matter can do. Functionalism in this broadest sense is so ubiquitous in science that it is tantamount to a reigning presumption of all of science. And since science is always looking for simplifications, looking for the greatest generality it can muster, functionalism in practice has a bias in favor of minimalism, of saying that less matters than one might have thought. The law of gravity says that it doesn't matter what stuff a thing is made of—only its mass matters (and its density, except in a vacuum). The trajectory of cannonballs of equal mass and density is not affected by whether they are made of iron, copper or gold. It *might* have mattered, one imagines, but in fact it doesn't. And wings don't *have* to have feathers on them in order to power flight, and eyes don't have to be blue or brown in order to see. Every eye has many more properties than are needed for sight, and it is science's job to find the maximally general, maximally noncommittal—hence minimal—characterization of whatever power or capacity is under consideration. Not surprisingly, then, many of the disputes in normal science concern the issue of whether or not one school of thought has reached too far in its quest for generality.

Since the earliest days of cognitive science, there has been a particularly bold brand of functionalistic minimalism in contention, the idea that just as a heart is basically a pump, and could in principle be made of anything so long as it did the requisite pumping without damaging the blood, so a mind is fundamentally a control system, implemented in fact by the organic brain, but anything else that could *compute the same control functions* would serve as well. The actual matter of the brain—the chemistry of synapses, the role of calcium in the depolarization of nerve fibers, and so forth—is roughly as irrelevant as the chemical composition of those cannonballs. According to this tempting proposal, even the underlying microarchitecture of the brain's connections can be ignored for many purposes, at least for the time being, since it has been proven by computer scientists that any function that can be computed by one specific computational architecture can also be computed (perhaps much less efficiently) by another architecture. If all that matters is the computation, we can ignore the brain's wiring diagram, and its chemistry, and just worry about the "software" that runs on it. In short—and now we arrive at the provocative version that has caused so much misunderstanding—in principle you could replace your wet, organic brain with a bunch of silicon chips and wires and go right on thinking (and being conscious, and so forth).

This bold vision, computationalism or "strong AI" (Searle), is composed of two parts: the broad creed of functionalism—

handsome is as handsome does—and a specific set of minimalist empirical wagers: neuroanatomy doesn't matter; chemistry doesn't matter. This second theme excused many would-be cognitive scientists from educating themselves in these fields, for the same reason that economists are excused from knowing anything about the metallurgy of coinage, or the chemistry of the ink and paper used in bills of sale. This has been a good idea in many ways, but for fairly obvious reasons, it has not been a *politically* astute ideology, since it has threatened to relegate those scientists who devote their lives to functional neuroanatomy and neurochemistry, for instance, to relatively minor roles as electricians and plumbers in the grand project of explaining consciousness. Resenting this proposed demotion, they have fought back vigorously. The recent history of neuroscience can be seen as a series of triumphs for the lovers of detail. Yes, the specific geometry of the connectivity matters; yes, the location of specific neuromodulators and their effects matter; yes, the architecture matters; yes, the fine temporal rhythms of the spiking patterns matter, and so on. Many of the fond hopes of opportunistic minimalists have been dashed—they had hoped they could leave out various things, and they have learned that no, if you leave out x, or y, or z, you can't explain how the mind works.

This has left the mistaken impression in some quarters that the underlying idea of functionalism has been taking its lumps. Far from it. On the contrary, the reasons for accepting these new claims are precisely the reasons of functionalism. Neurochemistry matters because—and *only* because—we have discovered that the many different neuromodulators and other chemical messengers that diffuse through the brain have *functional roles* that make important differences. What those molecules *do* turns out to be important to the *computational* roles played by the neurons, so we have to pay attention to them after all. To see what is at stake here, compare the neuromodulators to the food that is ingested by people. Psychologists and neuroscientists do not, as a rule, carefully inventory the food intake of their subjects, on the entirely plausible grounds that a serving of vanilla ice cream makes roughly the same contribution to how the brain goes about its tasks as a serving of strawberry ice cream. So long as there isn't any marijuana in the brownies, we can ignore the specifics of the food, and just treat it as a reliable energy source, the brain's power supply. This *could* turn out to be mistaken. It might turn out that psychologically important, if subtle, differences, hinged on whether one's subjects had recently had vanilla ice cream. Those who thought it did make a difference would have a significant empirical disagreement with those who thought it didn't, but this would not be a disagreement between functionalists and antifunctionalists. It would be a disagreement between those who thought that functionalism had to be expanded downward to include the chemistry of food and those who thought that functionalism could finesse that complication.

Consider the following:

> there may be various general neurochemical dispositions [based on the neuropeptide systems] that guide the patterning of thoughts that no amount of computational work can clarify. (Panskepp 1998, p. 36)

This perfectly captures a widespread (and passionately endorsed) attitude, but note that there is nothing oxymoronic about a computational theory of neuromodulator diffusion and its effects, for instance, and pioneering work in "virtual neuromodulators" and "diffusion models of computational control" is well underway. Minds will turn out not to be *simple* computers, and their computational resources will be seen to reach down into the subcellular molecular resources available only to organic brains, but the theories that emerge will still be functionalist in the broad sense.

So *within* functionalism broadly conceived a variety of important controversies have been usefully playing themselves out, but an intermittently amusing side effect has been that many neuroscientists and psychologists who are rabidly anti-computer and anti-AI for various ideological reasons have mistakenly thought that philosophers' *qualia* and *zombies* and *inverted spectra* were useful weapons in their battles. So unquestioning have they been in their allegiance to the broad, bland functionalism of normal science, however, that they simply haven't imagined that philosophers were saying what those philosophers were actually saying. Some neuroscientists have befriended *qualia*, confident that this was a term for the sort of functionally characterizable complication that confounds oversimplified versions of computationalism. Others have thought that when philosophers were comparing zombies with conscious people, they were noting the importance of emotional state, or neuromodulator imbalance. I have spent more time than I would like explaining to various scientists that their controversies and the philosophers' controversies are not translations of each other as they had thought but false friends, mutually irrelevant to each other. The principle of charity continues to bedevil this issue, however, and many scientists generously persist in refusing to believe that philosophers can be making a fuss about such a narrow and fantastical division of opinion.

Meanwhile, some philosophers have misappropriated those same controversies within cognitive science to support their claim that the tide is turning against functionalism, in favor of qualia, in favor of the irreducibility of the "first-person point of view" and so forth. This widespread conviction is an artifact of interdisciplinary miscommunication and nothing else.

5 The Future of an Illusion

I do not know how long this ubiquitous misunderstanding will persist, but I am still optimistic enough to suppose that some time in the next century people will look back on this era and marvel at the potency of the visceral resistance[18] to the obvious verdict about the Zombic Hunch: it is an illusion.

18. It is visceral in the sense of being almost entirely arational, insensitive to argument or the lack thereof. Probably the first to comment explicitly on this strange lapse from reason among philosophers was Lycan, in a footnote at the end of his 1987 book, *Consciousness*, that deserves quoting in full:

> On a number of occasions when I have delivered bits of this book as talks or lectures, one or another member of the audience has kindly praised my argumentative adroitness, dialectical skill, etc., but added that cleverness—and my arguments themselves—are quite beside the point, a mere exercise and/or display. Nagel (1979 [preface to *Mortal Questions*, Cambridge Univ. Press]) may perhaps be read more charitably, but not much more charitably:
>
>> I believe one should trust problems over solutions, intuition over arguments. ... [Well, excuuuuuse me!—WGL] If arguments or systematic theoretical considerations lead to results that seem intuitively not to make sense ... , then something is wrong with the argument and more work needs to be done. Often the problem has to be reformulated, because an adequate answer to the original formulation fails to make the *sense* of the problem disappear. (pp. x–xi)

Will the Zombic Hunch itself go extinct? I expect not. It will not survive in its current, toxic form but will persist as a less virulent mutation, still psychologically powerful but stripped of authority. We've seen this happen before. It still *seems* as if the Earth stands still and the Sun and Moon go around it, but we have learned that it is wise to disregard this potent appearance as mere appearance. It still *seems* as if there's a difference between a thing at absolute rest and a thing that is merely not accelerating within an inertial frame, but we have learned not to trust this feeling. I anticipate a day when philosophers and scientists and laypersons will chuckle over the fossil traces of our earlier bafflement about consciousness: "It still *seems* as if these mechanistic theories of consciousness leave something out, but of course that's an illusion. They do, in fact, explain everything about consciousness that needs explanation."

If you find my prediction incredible, you might reflect on whether your incredulity is based on anything more than your current susceptibility to the Zombic Hunch. If you are patient and open minded, it will pass.

If by this Nagel means only that intuitions contrary to ostensibly sound argument need at least to be explained away, no one would disagree (but the clause "something is wrong with the argument" discourages that interpretation). The task of explaining away "qualia"-based intuitive objections to materialism is what in large part I have undertaken in this book. If I have failed, I would like to be *shown why* (or, of course, presented with some new antimaterialist argument). To engage in further muttering and posturing would be idle. (1987, pp. 147–148)

Consciousness is often celebrated as a mystery beyond science, impenetrable from the outside, however intimately known to each of us from the inside. I think this tradition is not just a mistake, but a serious obstacle to ongoing scientific research that *can* explain consciousness, just as deeply and completely as it can explain other natural phenomena: metabolism, reproduction, continental drift, light, gravity, and so on. In order to lay—or clarify—the foundations for this research, I will begin by considering claims about the supposed limits of all such scientific investigations of consciousness.

1 Scientists from Mars

Suppose scientifically and technologically advanced "Martians" came to Earth to study the fauna and flora here. Let's assume them to have some kind of sense organs, which might be as different from human senses as you can imagine, so long as these permit them to acquire information about physical regularities in the world about as readily as we can. Being technologically advanced, they can thus do what we have done with microscopes, telescopes, infrared and ultraviolet detectors, chemical "sniffers" and the like: they can arrange to "see" what we can see, "hear" what we can hear, and so forth, thanks to prosthetic extensions of whatever senses they have, suitably equipped with Martian user-interfaces. Then whatever is observable to us is observable to them, albeit indirectly on some occasions (the way the shapes of bacteria, the shadows cast by infrared sources of electromagnetic radiation, and the vibrations emitted by distant earthquakes are observable by us thanks to our devices).

Among the phenomena that would be readily observable by these Martians would be all our *public* representations of consciousness: cartoon "thought balloons" such as the Steinberg masterpiece on the cover, soliloquies in plays, voice-overs in films, use of the *omniscient author* point of view in novels, and so forth. We tend to overlook the fact that much of what "we"— you and I and our friends and neighbors—believe about consciousness comes from our huge supply of shared, public, objective representations of the streams of consciousness of other folks, real or fictional. They would also have available to them the less entertaining representations of consciousness found in all the books by philosophers, psychologists, neuroscientists, phenomenologists, and other sober investigators of the phenomena. From all of this the anthropologists among them (the exomartian faunologists) would be able to arrive at an elaborate account of that part of the behavior of *H. sapiens* (as we communicating Earthlings call ourselves) that concerns the *folk theory of consciousness* as well as our early stabs at a scientific theory of consciousness.

Digression: I am supposing that these Martians already have the knack of adopting the *intentional stance* (Dennett 1971, 1987) toward the fauna they observe, so that they can learn our languages and interpret our public communication, but I am not presupposing that these Martians are themselves conscious in any of the tendentious ways much discussed of late by philosophers. So for the sake of argument consider that the Martians might be *zombies*, whose data-gathering and scientific theorizing is all accomplished without a trace of "phenomenality" or "qualia" or whatever you take to be the hallmark of real consciousness. That is, for the time being, I am not supposing that their manifest scientific expertise would be well-nigh conclusive evidence that they are conscious. They might, moreover, be quite unmoved by our music, our art, our theater, while unproblematically able to detect how much it matters *to us*. ("What *do* they see in these Picassos?" they ask incredulously, while noting not only the high prices we are willing to pay for them, but the large neuromodulator, endocrine, and visceral effects produced by encounters with them.)

My introduction of these imaginary Martian scientists permits me to expose and render vivid a familiar subliminal theme in the current debates about a scientific theory of consciousness. One of the tenets of the folk theory that the Martians would soon discover is that a scientific theory of consciousness is widely held by Earthlings to be impossible. Part of the lore that they would pick up—just as we pick it up, in the course of our enculturation—is that consciousness is utterly private, inaccessible to outsiders, somehow at least partly incommunicable and uninvestigatable by science—that is, by the very methods the Martians are using to explore our planet. Would they credit this? Would they understand it? Could they explain it? And, more pointedly, what would they make of the hypothesis that there was something that they, the Martians, couldn't know about human consciousness that we, the Earthlings, can know? They read Nagel's (1974) "What Is It Like to Be a Bat?" and thereby confront the question: "What is it like to be a human being?" They read David Chalmers's (1996) *The Problem of Consciousness* and wonder if they have even identified "the Hard Problem" of consciousness. What, if anything, about "our" consciousness is off-limits to these alien investigators? And if there is such a thing, how do "we" know that it is real?

One of the texts that the Martians would surely study is Descartes's *Meditations* (1641) and they would find it speaking quite forthrightly to them. The preface is directed to those who are willing "to meditate seriously *with me* and to lead the mind simultaneously away from the senses and away from all prejudices." Descartes would expect his Martian readers to perform for themselves the thought experiments and inferences, and to discount any peculiarities of their sensory apparatus ("away from the senses . . . away from all prejudices"). Good scientist that Descartes was, he appreciated the value of intersubjectivity, and the ways science has of canceling out the idiosyncrasies of individual investigators so that all can participate together in a shared inquiry, the "third-person" approach of scientific method. Martians are certainly not disqualified from joining in Descartes's meditations, and I propose that we follow Descartes's lead and strive for a maximally intersubjective science of consciousness. Let us see what happens when we try to cantilever this third-person methodology of science as far into the private interiors of minds as it will go. Will we leave important recesses untouched?

There is a considerable chorus of opinion these days insisting that these efforts must fail, that a purely third-person science of consciousness is methodologically impoverished, cut off from important sources of evidence, or data, or enlightenment . . . or something. We need, it is said, a "first-person science of consciousness" or even a "second-person science of consciousness" (which stresses empathy, and might be more properly be called the second-person *familiar* methodology—*du* not *Sie*, *tu* not *vous*). The idea, variously expressed or just tacitly presupposed, is that the Martians can't play these games that we can play. They can't engage in a first-person science of consciousness because they aren't themselves the right kind of first-persons. They can study *Martian* consciousness from the first-person point of view, if there is any, but not *our* consciousness. Or it might be said that they can't engage in a second-person science of consciousness because, being an alien life form, they can't form the *I–thou* bond of empathy such a method presupposes.

My question is this: Is there any good reason to believe any of this? And my answer will be No, there is nothing about our consciousness that is inaccessible to the maybe metal-headed methods of the Martians. The third-person methods of the natural sciences suffice to investigate consciousness as completely as any phenomenon in nature can be investigated, without *significant* residue. What is the import of "significant" here? Simply this: If scientists were to study a single grain of sand, there would always be more that could be discovered about it, no matter how long they worked. The sums of the attractive and repulsive forces between all the subatomic particles composing the atoms composing the grain will always have some residual uncertainty in the last significant digit we have calculated to date, and backtracking the location in spacetime of the grain of sand over the eons will lead to a spreading cone of indiscernibility. But our ignorance will not be significant. The principle of diminishing returns applies. My claim is that if we use the third-person methods of science to study human consciousness, whatever residual ignorance we must acknowledge "at the end of the day" will be no more unsettling, no more frustrating or mystifying, than the ignorance that is inelimininable when we study photosynthesis, earthquakes, or grains of sand. In short, no good reasons have been advanced for the popular hypothesis that consciousness is, from the point of view of third-person science, a *mystery* in a way that other natural phenomena are not. Nor are there good reasons for claiming that there is something significant that we know (or you know) by being consciousness that is utterly beyond the ken of the Martian scientists, however different they are from us.

We can begin to approach this issue by asking some boundary-setting questions. If Martians trying to study human consciousness must perforce leave something out, how do *we* know this? And who is this "we"? Is there something Francophones know about their consciousness that others cannot know? Is there something women know about women's consciousness that men can never know? Do right-handers know things about right-hander consciousness that left-handers can never know? Is there something *you* know about your own consciousness that we others can never know? Nagel's classic paper gently resists— without supporting argument—this retreat into solipsism, suggesting that "we" could know what it is like for "us" to experience things the human way, while insisting—without supporting argument—that we could not know what it is like to be something as different as a bat. The reason, I submit, that Nagel could help himself to this denial of solipsism is simply that nobody wants to challenge it; it appeals to people. It appeals to *us*. We—nudge, nudge—know about our consciousness because we communicate about it all the time. In our everyday dealings with each other we *presuppose* a vast sharing of understanding in all our public representations of consciousness, and as we contribute to that common stockpile, our presupposition is apparently vindicated.

The folk theory of human consciousness is a hugely successful mutual enterprise, but it does have its well-known puzzlepoints. Can a person born blind share "our" understanding of color? What about a color-blind person? What about "spectrum inversion," a thought experiment at least three hundred years old? Might it be that what I see as blue you see as yellow, but nevertheless you call that subjective color blue? What is it like to be an infant—is it a "buzzing blooming confusion" or something very different from that? Do men and women actually experience the world in ways that are fundamentally incomparable? There are lots of competing answers to these puzzle questions, and they all deserve our attention eventually, but rather than trying to adjudicate them *from the outset*, we should take a deep breath and recognize that all the answers, good and bad, are themselves parts of *folk theories* of consciousness, not *data* that we can share with the Martians.

2 Folk Theories and Philosophy

Back in the 1970s, the AI researcher Patrick Hayes (1978) embarked on a project to formalize a portion of folk physics or what he called *naive* physics—the physics we all betray knowledge of in our everyday life: towels absorb water, shadows can be projected through clear glass, things drop when you let go of them and often bounce (depending on the surface they land on), when things collide they make a noise, and so forth. We literally couldn't live without naive physics; it is extremely swift and fecund in its deliverance of reliable expectations, and virtually involuntary. You can't readily turn off your expectations. For instance, you "unthinkingly" leap back from the table when a water glass is overturned, expecting the water to roll off the edge onto your lap. *Somehow* your brain generated that expectation from its current perceptual cues and took presumably appropriate avoidance action. The background machinery of naive physics is not directly accessible to introspection, but can be studied indirectly by mapping its "theorems," the generalizations it can be seen to endorse (in a manner of speaking) by its particular deliverances. Many magic tricks exploit our intuitions of naive physics, gulling us into overlooking "impossible" possibilities, or inducing us to jump to conclusions (unconsciously) on the basis of a perceptual cue of one sort or another. Then there are the counterintuitive phenomena that baffle us naive physicists: gyroscopes, pipettes (why on earth doesn't the Pepsi fall out of the bottom of that straw—it's wide open!), siphons, sailing upwind, and more. Hayes's delicious idea was to try to formalize the naive physics of liquids, yielding a *theory* that would predict all the things we actually expect from liquids and hence predict *against* the things liquids do that we view as anomalies, such as siphons. Siphons are "physically impossible" according to naive physics.

What Hayes set out to do was a kind of rigorous anthropology, attempting to axiomatize the false theories found among the folk. Let's call it *aprioristic anthropology of naive physics*, to mark its resolute refusal to let the actual facts get in the way of deducing the implications of its found axioms. The physics was naive, but Hayes was not. His project was *sophisticated aprioristic anthropology*, since he was fully alert to the fact that false theories are just as amenable to formalization as true theories, and he withheld all allegiance to his axioms.

One could attempt just the same sort of project with folk psychology: deducing the implications of whatever is deemed "axiomatic" (unquestioned, impossible to deny, too obvious for words) by the folk. Call this enterprise *sophisticated aprioristic anthropology of folk (naive) psychology*. The theory educed should, like naive physics, rule out as flat impossible whatever psychological anomalies draw a stare of disbelief or utter bafflement from the folk. (*"That can't happen!"*) So blindsight, blindness denial, prosopagnosia (inability to recognize faces), and a variety of other well-known bizarre phenomena—the pipettes and gyroscopes of psychology—should have theorems of formal folk psychology denying them. It is tempting to interpret the field of philosophy of mind as just this endeavor: an attempt at a rigorous unification and formalization of the fundamental intuitions the folk manifest in both their daily affairs and in reflective interaction with the questioning anthropologists. "Consult your intuitions," say the philosophers. "Do they agree with the following proposition? . . ." And if the task were done well, it would yield a valuable artifact for further study: the optimized "theory" of late-twentieth-century-Anglophone folk psychology. It could be compared to similar refinements of the folk theories of other peoples, speaking other languages, at other times and places. It hardly needs saying that such a research tradition would have a lot in common with the attempts by linguists to codify, formally, the grammars of natural languages, yielding all and only the grammatical sentence—that is to say, the sentences that sound all right, on reflection, to native speakers. (*"You can't say that in English!"*) These are investigations worth doing, and the results are bountifully illuminating.

But although this interpretation would give philosophers of mind a clear and valuable job description consistent with much of what they have been doing for the last half century, in one regard it is a distortion, because the philosophers have taken themselves and their colleagues and students to be the folk under examination—*auto*anthropology—and then many of them have neglected to bracket their allegiance to the axioms they have uncovered. (Linguists have long confronted the epistemological problems of distortion that arise from consulting only their own native intuitions of grammaticality. Strangely, philosophers have not always been so self-critical in their methodological reflections.) Call the philosophers' enterprise *naive aprioristic autoanthropology*. They have proceeded as if the deliverances of their brute intuitions were not just *axiomatic-for-the-sake-of-the-project* but *true*, and moreover, somehow inviolable. (See Lycan's exposure of this theme in Nagel in the previous chapter, fn. 18.) One vivid but not always reliable sign of this is the curious reversal of the valence of the epithet "counterintuitive" among philosophers of mind. In most sciences, there are few findings more prized than a counterintuitive result. It shows something surprising and forces us to reconsider our often tacit assumptions. In philosophy of mind, a counterintuitive "result" (e.g., a mind-boggling implication of somebody's "theory" of perception, memory, consciousness, or whatever) is typically taken as tantamount to a refutation. This affection for one's current intuitions, sometimes amounting (as we saw in the previous chapter) to a refusal even to consider alternative perspectives, installs deep conservatism in the methods of philosophers. Conservatism can be a good thing, but only if it is acknowledged. By all means, let's not abandon perfectly good and familiar intuitions without a fight, but let's recognize that the intuitions that are initially used to frame the issues may not live to settle the issues.

If we let the elaboration of naive physics and academic physics be our guide, we can expect that in due course much of what is intuitive in folk psychology will be vindicated, incorporated, explained by our advanced academic theories of the phenomena. After all, the reason naive physics is so valuable to us is that its deliverances are, in the main, true. We should expect folk psychology to be similarly rich in retrospectively confirmed truth. But we can't count on it.

3 Heterophenomenology Revisited

I propose, then, that we follow Descartes's lead, and start with the data that we know we share with the Martians, and see where it leads us. Among the data are facts about which folk theories we Earthlings (or we Anglophone philosophers of mind) hold. And one of the open research questions before us is: which of our folk theorems will prove to be correct? Thus we do not presuppose at the outset that we won't discover, in the process of developing our scientific theory, that some folk theory is *right* about the inaccessibility of human consciousness to Martians. We just require that the case for such a discovery be itself intersubjectively accessible. (And this must be presupposed by those who argue for such claims—else why are they wasting their breath and our time?) We can readily imagine, at least superficially, how this would go: the Martians acknowledge ever growing frustration in their attempts to predict, anticipate, account for phenomena that the Earthling scientists are making real progress on. Attempts by the Earthlings to teach the Martians how to proceed not only fail but show telltale signs of systematic contrariness (recall that favorite factoid of physics, Heisenbergian uncertainty). We eventually come to see that it's not just that these Martians have a tin ear for Earthling phenomena of consciousness; the tin ear is uncorrectable by any imaginable prosthesis or training. *And* we can explain why.

Nagel offers us a reason, but it begs the question: "If the subjective character of experience is fully comprehensible only from one point of view, then any shift to greater objectivity—that is, less attachment to a specific viewpoint—does not take us nearer to the real nature of the phenomenon: It takes us farther away from it" (1974, p. 447). But the antecedent of his conditional has not yet been shown. It may seem too obvious to need a demonstration, but if so, then even a rudimentary attempt to deny it should expose itself to decisive refutation. Let's see. The third-person method, the method both we and Martians can adopt and know we have adopted in common, is captured in the strictures of what I have dubbed *heterophenomenology* (Dennett 1982, 1991):

> the *neutral* path leading from objective physical science and its insistence on the third-person point of view, to a method of phenomenological description that can (in principle) do justice to the most private and ineffable subjective experiences, while never abandoning the methodological principles of science. (1991, p. 72)

There is nothing revolutionary or novel about heterophenomenology; it has been practiced, with varying degrees of punctiliousness about its presuppositions and prohibitions, for a hundred years or so, in the various branches of experimental psychology, psychophysics, neurophysiology, and today's cognitive neuroscience. I just gave it a name and got particularly self-conscious about identifying and motivating its enabling assumptions.

We heterophenomenologists start with *recorded raw data* on all the physical goings-on inside and outside our subjects, a pool restricted to communicating human beings (with or without identifiable pathologies and quirks, of both sexes, of all ages, cultures, varying socioeconomic status, etc., etc.). Note that our Martian collaborators are excluded from the subject pool but are deemed at the outset to be fully qualified to be fellow investigators. We gather data on all the chemical, electrical, hormonal, acoustical . . . and other physical events occurring in the subjects, and we pay particular attention to the timing of all these events, but we also single out one data stream from the others for special treatment. We take some of the noises and marks made by subjects as consisting of communication—oral and otherwise—and compose transcripts, which then are further interpreted to yield an inventory of speech acts, which are further interpreted as (apparent) expressions of belief.

This transformation of the raw data of acoustic pressure waves, lip-movements, button-pressings and such into expressions of belief *requires* adopting the intentional stance. It requires us to treat the subjects as if they were believers and desirers capable of framing and executing speech acts with intended meanings—but it leaves wide open the vexatious question, from folk theory of consciousness, of whether or not some subjects might be zombies. It also leaves untouched such subordinate puzzles of folk theory as whether zombies should be properly said to perform *real* speech acts or merely *apparent* speech acts, and whether zombies thereby express their beliefs or merely *seem* to express their *apparent* beliefs, and so forth. Some people believe that zombie hypotheses are serious problems; for them, these are serious questions, but they don't have to be settled *ab initio*. We can conveniently postpone them, noting that it is agreed on all sides that the intentional stance works exactly the same for zombie behavior as for the behavior of genuinely conscious beings, supposing these to be different. By definition, philosophical zombies are behaviorally indistinguishable from conscious beings, and the intentional stance is *behavioristic* in the sense of restricting itself to the intersubjectively observable "behavior" of all the subjects, and all their parts, internal and external. It is not behavioristic in another sense, of course, since it precisely consists in "mentalistic" or "intentionalistic" interpretations of raw behaviors, identifying them as actions, expressive of beliefs, desires, intentions, and other propositional attitudes.

Is this neutrality of the intentional stance on the zombie problem a bug or a feature? From the vantage point of our attempt to found a natural science of human consciousness, it is most definitely a feature; it is what permits us to postpone the perplexities of folk theory while getting on with the business of extracting, organizing, and interpreting the data we and the Martians share. What is it like to be a zombie? By definition: nothing. But even those who take zombies seriously agree that there *seems* (at least to us, on the outside) to be something it is like to be a zombie, and just this seeming is what heterophenomenology scrupulously captures:

> In this chapter we have developed a *neutral* method for investigating and describing phenomenology. It involves extracting and purifying *texts* from (apparently) speaking *subjects*, and using those texts to generate a theorist's fiction, the subject's *heterophenomenological world*. This fictional world is populated with all the images, events, sounds, smells, hunches, presentiments, and feelings that the subject (apparently) sincerely believes to exist in his or her (or its) stream of consciousness. Maximally extended, it is a neutral portrayal of exactly *what it is like to be* that subject—in the subject's own terms, given the best interpretation we can muster. People undoubtedly do believe

that they have mental images, pains, perceptual experiences, and all the rest, and *these* facts—the facts about what people believe, and report when they express their beliefs—are phenomena any scientific theory of the mind must account for. (Dennett 1991, p. 98)

Working side by side, we and the Martians move from *raw* data to *interpreted* data: convictions, beliefs, attitudes, emotional reactions . . . but all these are *bracketed* for neutrality. Why bracket? Because of two possible failures of overlap, familiar from the judicial injunction to tell the whole truth and nothing but the truth: subjects often fail to tell the whole truth because some of the psychological things that happen in them are unsuspected by them, and hence go unreported, and subjects often fail to tell nothing but the truth because they are tempted into theorizing that goes beyond what we can demonstrate to be the limit of their experience. Bracketing has the effect of holding them to an account of *how it seems to them* without pre-judging, for or against, the questions of whether how it seems to them is just how it is.

Consider, for instance, the well-studied phenomenon of *masked priming*. It has been demonstrated in hundreds of different experiments that if you present subjects with a "priming" stimulus, such as a word or picture flashed briefly on a screen in front of the subject, followed very swiftly by a "mask"— a blank or sometimes randomly patterned rectangle—before presenting the subjects with a "target" stimulus to identify or otherwise respond to, there are conditions under which subjects will manifest behavior that shows they have discriminated the priming stimulus, while they candidly and sincerely report that they were entirely unaware of any such stimulus. For instance, asked to complete the word stem *fri__*, subjects who have been shown the (masked) priming stimulus *cold* are more likely to comply with *frigid* and subjects who have been shown the priming stimulus *scared* are more likely to comply with *fright* or *frightened*, even though both groups of subjects claim not to have seen anything but first a blank rectangle followed by the target to be completed. Now are subjects to be trusted when they say that they were not conscious of the priming stimulus? There are apparently two ways theory can go here:

A. Subjects are conscious of the priming stimulus and then the mask makes them immediately forget this conscious experience, but it nevertheless influences their later performance on the target.

B. Subjects unconsciously extract information from the priming stimulus, which is prevented from "reaching consciousness" by the mask.

It is open for scientific investigation to develop reasons for preferring one of these theoretical paths to the other, but at the outset, heterophenomenology is neutral, leaving the subjects' heterophenomenological worlds bereft of any priming stimuli—that is how it seems to the subjects, after all—while postponing an answer to the question of how or why it seems thus to the subjects.

Heterophenomenology is the beginning of a science of consciousness, not the end. It is the organization of the data, a catalog of *what must be explained*, not itself an explanation. And in maintaining this neutrality, it is actually doing justice to the *first-person* perspective, because you yourself, as a subject in a masked priming experiment, cannot discover anything in your experience that favors A or B. (If you think you can discover something—if you notice some glimmer of a hint in the experience, speak up! You're the subject, and you're supposed to tell it like it is! Don't mislead the experimenters by concealing something you discover in your experience. Maybe they've set the timing wrong for you. Let them know. But if they've done the experiment right, and you really find, so far as you can tell from your own first-person perspective, that you were not conscious of any priming stimulus, then say so, and note that both A and B are still options between which you are powerless to offer any further evidence.)

In other phenomena, what needs to be bracketed is subjects' manifestly false beliefs about what is present in their own experience. For instance, most people—"naive subjects" in the standard jargon—suppose that their color vision extends all the way to the periphery of their visual fields, and they also suppose that their visual fields are approximately as detailed or fine-grained all the way out. They are astonished when it is demonstrated to them that they cannot identify a playing card—cannot even say if it is red or black—even though they can see it being wiggled at the edge of their visual fields. Motion detection extends well beyond color vision in our visual fields, and this is just one of the incontestable facts that play havoc with the folk psychology of vision. What needs to be explained by a science of consciousness in this case is the etiology of a false belief. The question to ask, and answer, is

Why do people *think* that their visual fields are detailed and colored all the way out?

not:

Why, since people's visual fields *are* detailed and colored all the way out (that's what they tell us), can't they identify things they see moving in the parafoveal parts of their visual fields?

There is an amiable but misleading tendency of people to exaggerate the wonders of their own conscious experience, rather like audiences at stage magic shows, who tend to leave the theater claiming to have witnessed more marvels than were actually presented for their enjoyment. So the astringent neutrality of heterophenomenology often has the deflationary effect of cutting the task of explaining consciousness down to size; consciousness is not quite as supercalifragilisticexpialidocious as many people like to believe. But the goal of heterophenomenology is getting at the data, whatever they are, not deflation.

I have just noted that the neutrality of heterophenomenology actually does justice to first-person experience, a point often overlooked by its critics. This is partly the result of misdirection not unlike the confusion sown by the different senses of the term "behaviorist." Consider the following passage from a recent paper by Parvizi and Damasio, commenting on a shift in perspective in the maturing of cognitive neuroscience. They disparage

> a time in which the phenomena of consciousness were conceptualized in exclusively behavioral, third-person terms. Little consideration was given to the cognitive, first-person description of the phenomena, that is, to the experience of the subject who is conscious. (Parvizi and Damasio 2001, p. 136)

Notice that the new, improved perspective gives consideration to "the cognitive, first-person *description* of the phenomena"—in short, the new improved perspective *is* heterophenomenology. What it is being contrasted with is an old-fashioned behavioristic (in the anti-intentionalist sense) abstemiousness that refused to consider subject's descriptions as anything other than noise-producing behavior.

The fossil traces of this bias against the *serious use of the intentional stance* are still blunting the tools of cognitive science. We prepare our subjects with a very carefully worded set of instructions, and debrief them at the end of the experiment to make sure they were following instructions, and this provides some obligatory quality control when we go to interpret their button presses as speech acts, as well as an opportunity to uncover unsuspected sources of data contamination, but we still tend to *minimize* the use of subject–experimenter communication, treating it as more trouble than it is worth. After the experimenter has squeezed out as much of the variation in subject performance as possible, the residual individuality of subjects is treated as a problem, not an opportunity, in most experimental settings. Alan Kingstone (pers. comm.) has recently observed that many of the research paradigms in cognitive science working on attention squander valuable opportunities by insisting on treating the variation in subject performance as noise to be statistically overcome instead of as invaluable signs of variations in subject's evanescent attitudes, idiosyncratic methods, lapses in attention, and the like. Designing experiments to exploit this variability is still a relatively rare practice, but nothing in the principles of heterophenomenology discourages it. On the contrary, it has always been prized when it could be exploited, however opportunistically, in special settings. Penfield's (1958) famous probes by cortical stimulation and query of awake patients were done fifty years ago. As our new neuro-imaging technology makes possible ever finer-grained probing under relatively noninvasive circumstances, we have only just begun exploring these avenues systematically. And there is still plenty of work to be done outside the brain-scanner. (Indeed, I would venture the opinion that innovations in experimental design that are quite independent of brain imaging technology will be the main source of discovery in the next few decades.)

Heterophenomenology is an inclusive methodology, and many of its branches have hardly been explored. As chapter 7 will note in more detail, the special issue of *Cognition* (2001) devoted to the cognitive neuroscience of consciousness in which Parvizi and Damasio's paper appears recounts a wide variety of recent work in many laboratories, and all of it is conducted according to the constraints of heterophenomenology. As the researchers insist, this methodological restraint does not prevent the research from taking the first-person point of view seriously.

A philosopher who has criticized heterophenomenology's neutrality is Joseph Levine (1994), who has claimed that "conscious experiences themselves, not merely our verbal judgments about them, are the primary data to which a theory must answer" (p. 117). Levine's claim can be most clearly understood in terms of a nesting of proximal sources that are presupposed as we work our way from raw data to heterophenomenological worlds:

(a) "conscious experiences themselves";

(b) beliefs about these experiences;

(c) "verbal judgments" expressing those beliefs;

(d) utterances of one sort or another.

What are the "primary data"? For heterophenomenologists, the *primary* data are the sounds recorded when the subjects' mouths move, or (d) the utterances, the *raw* uninterpreted data. But before we get to theory, we can interpret these data, carrying us via (c) speech acts to (b) beliefs about experiences. These are the primary *interpreted* data, the pretheoretical data, the *quod erat explicatum* (as organized into heterophenomenological worlds), for a science of consciousness. In his quest for primary data, Levine wants us to go all the way to (a) conscious experiences themselves, instead of stopping with (b) subjects' beliefs about their experiences, but this is not a good idea. If (a) outruns (b)— if you have conscious experiences you don't believe you have— those extra conscious experiences are just as inaccessible *to you* as to the external observers. So a first-person approach garners you no more usable data than heterophenomenology does. Moreover, if (b) outruns (a)—if you believe you have conscious experiences that you don't in fact have—then it is your beliefs that we need to explain, not the nonexistent experiences! Sticking to the heterophenomenological standard, and treating (b) as the maximal set of primary data, is a good way of avoiding a commitment to spurious data.

But what if some of your beliefs are inexpressible in verbal judgments? If you believe *that*, you can tell us, and we can add that belief to the list of beliefs in our primary data:

S claims that he has ineffable beliefs about *X*.

If this belief is true, then we encounter the obligation to explain what these beliefs are and why they are ineffable. If this belief is false, we still have to explain why *S* believes (falsely) that there are these particular ineffable beliefs. As I put it in *Consciousness Explained*:

> You are *not* authoritative about what is happening in you, but only about what *seems* to be happening in you, and we are giving you total, dictatorial authority over the account of how it seems to you, about *what it is like to be you*. And if you complain that some parts of how it seems to you are ineffable, we heterophenomenologists will grant that too. What better grounds could we have for believing that you are unable to describe something than that (1) you don't describe

> it, and (2) confess that you cannot? Of course you might be lying, but we'll give you the benefit of the doubt. (1991, pp. 96–97)

Another philosopher who has challenged the neutrality of heterophenomenology is David Chalmers. I am claiming that heterophenomenology's resolutely third-person treatment of belief attribution squares perfectly with standard scientific method: when we assess the attributions of belief relied on by experimenters (in preparing and debriefing subjects, for instance) we use the principles of the intentional stance to settle what it is reasonable to postulate regarding the subjects' beliefs and desires. Now Chalmers has objected (in a debate at Northwestern University, February 15, 2001, from which this section is drawn) that this "behavioristic" treatment of belief is itself question-begging against an alternative vision of belief in which, for instance, "having a phenomenological belief doesn't involve just a pattern of responses, but often requires having certain experiences" (personal correspondence, Feb. 19, 2001). On the contrary, heterophenomenology is neutral on just this score. Surely we mustn't *assume* that Chalmers is right that there *is* a special category of "phenomenological" beliefs—that there is a kind of belief that is off-limits to zombies but not to us conscious folks. Heterophenomenology allows us to proceed with our catalog of a subject's beliefs leaving it open whether any or all of them are Chalmers-style phenomenological beliefs or mere zombie-beliefs. (More on this later.) In fact, heterophenomenology permits science to get on with the business of accounting for the patterns in all these subjective beliefs without stopping to settle this imponderable issue. And surely Chalmers must admit that the patterns in these beliefs are among the phenomena that any theory of consciousness must explain.

4   David Chalmers as Heterophenomenological Subject

Of course it *still* seems to many people that heterophenomenology must be leaving something out. That's the Zombic Hunch. How does heterophenomenology respond to this? Very straightforwardly: by *including* the Zombic Hunch among the heartfelt convictions any good theory of consciousness must explain. One of the things that it falls to a theory of consciousness to explain is *why some people are visited by the Zombic Hunch*. Chalmers is one such, so let's look more closely at the speech acts Chalmers has offered as a subject of heterophenomenological investigation.

Here is Chalmers's definition of a zombie (his zombie twin):

> Molecule for molecule identical to me, and identical in all the low-level properties postulated by a completed physics, but he lacks conscious experience entirely. . . . he is embedded in an identical environment. He will certainly be identical to me *functionally*; he will be processing the same sort of information, reacting in a similar way to inputs, with his internal configurations being modified appropriately and with indistinguishable behavior resulting. . . . he will be awake, able to report the contents of his internal states, able to focus attention in various places and so on. It is just that none of this functioning will be accompanied by any real conscious experience. There will be no phenomenal feel. There is nothing it is like to be a Zombie. . . . (1996, p. 95)

Notice that Chalmers allows that zombies have internal states with contents, which the zombie can report (sincerely, one presumes, believing them to be the truth); these internal states have contents, but not conscious contents, only pseudo-conscious contents. The Zombic Hunch, then, is Chalmers's conviction that he has just described a real problem. It *seems to him* that there is a problem of how to explain the difference between him and his zombie twin:

The justification for my belief that I am conscious lies not just in my cognitive mechanisms but also in *my direct evidence* [emphasis added]; the zombie lacks that evidence, so his mistake does not threaten the grounds for our beliefs. (One can also note that the zombie doesn't have the same beliefs as us, because of the role that experience plays in constituting the contents of those beliefs.) (Chalmers's Web site: Reply to Searle)

This speech act is curious, and when we set out to interpret it, we have to cast about for a charitable interpretation. How does Chalmers's justification *lie in* his "direct evidence"? Although he says the zombie *lacks* that evidence, nevertheless the zombie *believes* he has the evidence, just as Chalmers does. Chalmers and his zombie twin are heterophenomenological twins: when we interpret all the data we have, we end up attributing to them exactly the same heterophenomenological worlds. Chalmers fervently believes he himself is not a zombie. The zombie fervently believes he himself is not a zombie. Chalmers *believes* he gets his justification from his "direct evidence" of his consciousness. So does the zombie, of course.

The zombie has the conviction that he has direct evidence of his own consciousness, and that this direct evidence is his justification for his belief that he is conscious. Chalmers must maintain that the zombie's conviction is false. He says that the zombie *doesn't* have the same beliefs as us "because of the role that experience plays in constituting the contents of those beliefs," but I don't see how this can be so. Experience (in the special sense Chalmers has tried to introduce) plays no role in constituting the contents of those beliefs, since ex hypothesi, if experience (in this sense) were eliminated—if Chalmers were to be suddenly zombified—he would go right on saying what he says, insisting on what he now insists on, and so forth.[1] Even if his "phenomenological beliefs" suddenly ceased to be phenomenological beliefs, he would be none the wiser. It would not *seem to him* that his beliefs were no longer phenomenological.

But wait, I am forgetting my own method and arguing with a subject! As a good heterophenomenologist, I must grant Chalmers full license to his deeply held, sincerely expressed convictions and the heterophenomenological world they constitute. And then I must undertake the task of explaining the etiology of his beliefs. Perhaps Chalmers's beliefs about his experiences will turn out to be true, though how that prospect could emerge eludes me at this time. But I will remain neutral. Certainly we shouldn't give them incorrigible status. (He's not the Pope.) The fact that some subjects have the Zombic Hunch shouldn't be considered grounds for revolutionizing the science of consciousness.[2]

5 The Second-Person Point of View

Moving from the third-person to the first-person point of view is just asking for trouble; you get no data not already available

1. "I simply say that invoking consciousness is not necessary to explain actions; there will always be a physical explanation that does not invoke or imply consciousness. A better phrase would have been 'explanatorily superfluous', rather than 'explanatorily irrelevant.'" (Chalmers's second reply to Searle, on his Web site: http://www.u.arizona.edu/~chalmers/discussions.html.)

2. Chalmers seems to think that conducting surveys of his audiences, to see what proportion can be got to declare their allegiance to the Zombic Hunch, yields important data. Similar data-gathering would establish the falsehood of neo-Darwinian theory and the existence of an afterlife.

to all the rest of us from the third-person point of view, and you risk sending yourself off on wild goose chases trying to pin down conscious experiences that you only think you're having.

What about the second-person point of view? What people seem to have in mind by this suggestion is either some sort of *empathy*, or a sort of *trust* that is distinct from the admittedly weird, unnaturally noncommittal attitude adopted by heterophenomenology. Let's consider trust first. This neutrality or agnosticism has been criticized by Alvin Goldman. In "Science, Publicity, and Consciousness" (1997), he says that heterophenomenology is not, as I claim, the standard method of consciousness research, since researchers "rely substantially on subjects' introspective beliefs about their conscious experience (or lack thereof)" (p. 532). In personal correspondence (Feb. 21, 2001, available as part of my debate with Chalmers, on my Web site, at http://ase.tufts.edu/cogstud/papers/chalmersdeb3dft.htm), he puts the point this way:

> The objection lodged in my paper [Goldman 1997] to heterophenomenology is that what cognitive scientists *actually* do in this territory is not to practice agnosticism. Instead, they rely substantially on subjects' introspective beliefs (or reports). So my claim is that the heterophenomenological method is not an accurate description of what cognitive scientists (of consciousness) standardly do. Of course, you can say (and perhaps intended to say, but if so it wasn't entirely clear) that this is what scientists *should* do, not what they *do* do.

I certainly would play the role of reformer if it were necessary, but Goldman is simply mistaken; the adoption of agnosticism is so firmly built into scientific practice these days that it goes without saying, which is perhaps why he missed it. Consider, for instance, the decades-long controversy about mental imagery, starring Roger Shepard, Stephen Kosslyn, and Zenon Pylyshyn, among many others. It was initiated by the brilliant experiments by Shepard and his students, in which subjects were shown pairs of line drawings like the pair shown in figure 2.1, and asked to press one button if the figures were different views of the same object (rotated in space) and another button if they were of different objects.

Most subjects claimed to solve the problem by rotating one of the two figures in their "mind's eye" or imagination, to see if it could be superimposed on the other. Were subjects really doing the "mental rotation" they claimed to be doing? By varying the angular distance actually required to rotate the two figures into congruence, and timing the responses, Shepard was able to establish a remarkably regular linear relation between latency of response and angular displacement. Practiced subjects, he reported, are able to rotate such mental images at an angular velocity of roughly 60° per second (Shepard and Metzler 1971). This didn't settle the issue, since Pylyshyn and others were quick to compose alternative hyptheses that could account for this striking temporal relationship. Further studies were

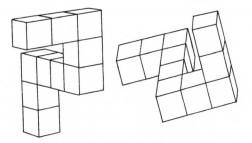

Figure 2.1
Shepherd blocks. Are these shapes the same or different?

called for and executed, and the controversy continues to generate new experiments and analysis today (see Pylyshyn 2002 for an excellent survey of the history of this debate, and also my commentary, 2002, both in *Behavioral and Brain Sciences*).

Subjects always *say* that they are rotating their mental images, so if agnosticism were not the tacit order of the day, Shepard and Kosslyn would never have needed to do their experiments to support subjects' claims that what they were doing (at least if described metaphorically) really was a process of image manipulation. Agnosticism is built into all good psychological research with human subjects. In psychophysics, for instance, the use of signal detection theory has been part of the canon since the 1960s, and it specifically commands researchers to control for the fact that the response criterion is under the subject's control although the subject is not him- or herself a reliable source on the topic. Or consider the voluminous research literature on illusions, both perceptual and cognitive, which standardly assumes that the data are what subjects judge to be the case, and never makes the mistake of "relying substantially on subjects' introspective beliefs."

The diagnosis of Goldman's error is particularly clear here: of course experimenters on illusions rely on subjects' introspective beliefs (as expressed in their judgments) about how things seem to them, but that *is* the agnosticism of heterophenomenology; to go beyond it would be, for instance, to assume that in size illusions there really are visual images of different sizes somewhere in subjects' brains (or minds), which of course no researcher would dream of doing.

Consider the illusion in figure 2.2. Does some traffic seem to be moving back and forth on the gray stripes? Nothing is moving on the page, but is that how it seems to you? Good.

Figure 2.2
Traffic, by Isia Leviant.

Now let's look in the brain and see what is happening in the visual cortex. Researchers would not expect to find patterns of excitation in the cortex that actually moved in synchrony with the apparent motion in your experience! They *might* find such a "movie in the brain," and that would be a truly revolutionary discovery; but the agnosticism of heterophenomenology excuses researchers from thinking that if they don't find such a movie, their subjects are lying to them. Their subjects are sincerely saying what it seems to them that they see—and this conviction is what needs explanation.

Finally, consider such phenomena as déjà vu. Sober research on this topic has never made the mistake of abandoning agnosticism about subjects' claims to be reliving previous experiences. See, for example, Bower and Clapper, in Posner 1989, or any good textbook on methods in cognitive science for the details.[3]

What about *empathy*, then. Is there some other sort of attitude, importantly different from the strange restraint of the heterophenomenological method, that might bear fruit in our quest for a scientific understanding of consciousness? Varela and Shear describe the empathy that they see as the distinguishing feature of a method they describe as first-person:

> In fact, that is how he sees his role: as an empathic resonator with experiences that are familiar to him and which find in himself a resonant chord. This empathic position is still partly heterophenomenological, since a modicum of critical distance and of critical evaluation is necessary, but the intention is entirely other: to meet on the same ground, as members of the same kind. . . . Such encounters would not be possible without the mediator being steeped in the

3. Goldman has responded to this paragraph in a series of e-mails to me, which I have included in an appendix available at http://ase.tufts.edu/cogstud/papers/chalmersdeb3dft.htm.

domain of experiences under examination, as nothing can replace that first-hand knowledge. This, then, is a radically different style of validation from the others we have discussed so far. (1999, p. 10)

One can hardly quarrel with the recommendation that the experimenter be "steeped in the domain of experiences" under examination, but, in a word, can Martians marinate? If not, why not? Is there more to empathy than just good, knowledgeable interpretation from the intentional stance? If so, what is it? In a supporting paper, Evan Thompson (2001) speaks of "sensual empathy," and opines: "Clearly, for this kind of sensual empathy to be possible, one's own body and the Other's body must be of a similar type." It may be clear to Thompson, but it is not clear to me. In fact, I think it is seriously mistaken.

It may, in the end, be true that "Martians" of some ilk would be incapable of sensual empathy with human beings, but this is hardly the sort of opinion on which a natural science of consciousness should be based. It should emerge, if it is true at all, from a discovered failure to connect, a striking disparity in the success of Martian and Earthling experimenters/investigators, for instance, and it should itself be a fact that our theory can explain, not something it presupposes in the very course of gathering its data. Any such gradient or discontinuity worth taking seriously can itself be discovered by heterophenomenology.

Note that Thompson's claim would enshrine Nagel's unargued assertion that "we" cannot know what it is like to be a bat as a methodological principle and render his claim off-limits to investigation. That can hardly be a good way for a science of consciousness to proceed, especially since some excellent work has already been done by "us" on what (if anything) it is like to be a bat! Akins (1993) shows how to proceed, and she reveals in the process that Nagel's assumption that *there is* a perspective or point of view occupied by a bat (a mustached bat, in her particular exploration) is nowhere near as secure as complacent philosophical tradition would suppose. If we don't want to legislate ourselves out of touch with such unsettling possibilities, we need to adopt a more neutral position. Instead of making it a methodological principle that aliens need not apply for positions on the research team, we should open the positions to all comers regardless of "body type" and see if this stands in the way.

My tentative, defeasible conclusion, then, is that my contention is so far unscathed: the method of heterophenomenology captures all the data for a theory of human consciousness in a neutral fashion. A "first-person" science of consciousness will either collapse into heterophenomenology after all, or else manifest an unacceptable bias in its initial assumptions.

It seems to many people that consciousness is a mystery, the most wonderful magic show imaginable, an unending series of special effects that defy explanation. I think that they are mistaken: consciousness is a physical, biological phenomenon—like metabolism or reproduction or self-repair—that is exquisitely ingenious in its operation, but not miraculous or even, in the end, mysterious.

Part of the problem of explaining consciousness is that there are powerful forces acting to make us think it is more marvelous than it actually is. As I noted in the previous chapter, in this regard consciousness resembles *stage* magic, a set of phenomena that exploit our gullibility, and even our desire to be fooled, bamboozled, awestruck. The task of explaining stage magic is in some regards a thankless task; the person who tells people how an effect is achieved is often resented, considered a spoilsport, a party pooper. I often get the impression that my attempts to explain some aspects of consciousness provoke similar resistance. Isn't it *nicer* if we all are allowed to wallow in the magical mysteriousness of it all? Or even this: if you actually manage to explain consciousness, they say, you will diminish us all, turn us into mere protein robots, mere *things*.

1 The Thankless Task of Explaining Magic

Such is the prevailing wind into which I must launch my efforts, but sometimes the difficulty of the task inspires strategies that exploit the very imagery that I wish in the end to combat. Lee Siegel draws our attention to the fundamental twist in his excellent book, *Net of Magic: Wonders and Deceptions in India*:

> "I'm writing a book on magic," I explain, and I'm asked, "Real magic?" By *real magic* people mean miracles, thaumaturgical acts, and supernatural powers. "No," I answer: "Conjuring tricks, not real magic." *Real magic*, in other words, refers to the magic that is not real, while the magic that is real, that can actually be done, is *not real magic*. (1991, p. 425)

It can't be *real* if it's explicable as a phenomenon achieved by a bag of ordinary tricks—cheap tricks, you might say. And that is just what many people claim about consciousness, too. So let's pursue the parallel with stage magic, and see how *some* of the effects of consciousness might be explained, and see how disappointing some of the explanations might be.

For more than a thousand years, the Indian Rope Trick has defied all attempts at explanation. I'm not alluding to some simple stunt in which a rope is thrown into the air, becomes rigid, and is then climbed by the agile magician. Many versions of that have been performed all over the world. I'm alluding to the *real* Indian Rope Trick, the Indian Rope Trick of legend, a much more shocking episode of magic:

> The magician throws a rope into the air, where it hangs, its top somehow invisible. A young assistant climbs the rope and disappears into thin air, but then is heard to taunt the magi-

cian, who takes a huge knife in his teeth and climbs the rope himself, disappearing in turn. A terrible fight is heard but not seen, and bloody limbs, a torso and a freshly severed head fall out of the sky onto the carpet beneath the rope. The magician reappears, climbing sadly down the rope, and bewailing the hot temper that has led him to murder his young assistant. He gathers up the bloody body parts and places them in a large covered basket, and asks the audience to join him in a prayer for the dead little boy, whereupon the lad jumps whole out of the basket, and all is well.

That is the heterophenomenological world of the Indian Rope Trick. That is what it would be like to witness the Indian Rope Trick. Has it ever been performed? Nobody knows. Thousands, probably millions, of people over hundreds of years have fervently believed that they themselves—or their brothers or uncles or cousins or friends—have witnessed the great spectacle with their own eyes. In 1875, Lord Northbrook offered the amazing fortune of 10,000 pounds sterling to anyone who could perform it. In the 1930s, the *Times of India* offered 10,000 rupees, and many others have offered huge rewards (Siegel 1991, pp. 199–200). The money has always gone unclaimed, so the sober judgment of those in the best position to know is that the Indian Rope Trick is a sort of archetypal urban legend, a mere intentional object, a notional trick, not a real one.

But wait: many people sincerely believe that the trick has been performed. Some of them, apparently, sincerely believe *that they have seen* the trick performed. If some people sincerely believe that they have seen the trick performed, doesn't that settle it? What else is a magic trick but the creation of sincerely held false beliefs about having witnessed one marvelous event or another? The magician doesn't *really* saw the lady in half; he only *makes you think you saw* him do it! If a magician can somehow or other make you think you saw him climb a rope, disappear, dismember a boy, and bring the boy back to life, he has performed the Indian Rope Trick, has he not? What more is required?

It matters how the belief is induced, it seems. If a magician managed somehow to hypnotize his entire audience, and then simply *told them* in gripping detail what he was doing, when he snapped his fingers and brought his audience awake with a standing ovation and exclamations of wonder, many of us would feel cheated. (This hypothesis of mass hypnotism—with or without hallucinogenic drugs covertly administered—has often been suggested, Siegel notes, by those who have sought to explain the birth of the legend.) Not that magic isn't always a bit of a cheat, but this is over the line, we feel. This doesn't count. It also doesn't count if the magician simply bribes people to *declare* they have seen the legendary feat after showing them some version of the simple rope-climbing stunt. In this case we have (c) verbal expression without (b) belief. But what if the effect of many such shills eagerly declaring their amazement managed to overwhelm some innocent audience members into sincerely avowing the same false belief?

Recall the famous demonstrations by Solomon Asch (1958) of the powers of group conformism. In a typical Asch experiment, "subjects" are asked which of three lines is the same length as a target line. Suppose line A is obviously the same length but a series of eight or nine "subjects"—confederates of Asch—all say "line B" before the one real subject in the experiment has to declare. Many subjects wilt under the social pressure and "agree" that line B is the match. When Asch debriefed his subjects, he found that while some of them admitted that they had been lured into expressing judgments they didn't believe, others insisted that by the time they were obliged to declare, they were convinced that their first impression had been wrong. They actually believed it when they expressed their hugely erroneous judgments. If a magician could create the Asch effect in a few subjects by using an army of shills, it would be *arguable* that the Indian Rope Trick had actually been performed on them, since they had been duped into believing they had witnessed the trick—that's how it seemed to them— but still I think we are inclined to consider such a method a cheat.

Coming by another, more high-tech route, if some magician with too much money commissioned the computer-graphics mavens at Industrial Light and Magic to create on videotape an ultrarealistic computer rendering of such a stunt, a videotape so apparently authentic that it could be sent as a "live feed" to CNN without the network's being able to determine that it was counterfeit, this, too, would be viewed by most if not all as not meeting the challenge to perform the trick. I doubt if you could collect the prize money with such a stunt, even though millions of people had been thereby convinced that they had seen a real event on "live" television. What is missing in these scenarios is actually not easy to say: it is quite all right to use smoke and mirrors, deceptive lighting, fake limbs and blood. Is it all right to use dozens of assistants? Yes, if they are backstage doing one thing or another, but what if they are disguised as audience members and are required to jump up and obscure the line of sight of the real audience members at crucial junctures? Where in the chain of causation leading to belief is the last permissible site of intervention? The "power of suggestion" is a potent tool in the magician's kit, and sometimes the magician's *words* play a more important role than anything the magician *shows* or *does*.

These observations draw our attention to the ill-behaved gaggle of tacit presumptions that govern our sense of what counts as a proper magic trick. It is not so embarrassing to acknowledge that our concept of *what counts* in such a case is in some regards disheveled, or unclear, since after all, we don't rest anything very heavy on our tacit understanding. Magicians may try to abuse our concept of magic, and all they risk is the loss of an audience if they misjudge what they can pass off as magic. It isn't brain science, after all. It's just show business.

But when the topic *is* brain science, something similar can take place. When we think about the phenomena of consciousness and wonder how they are accomplished in the brain, it is not at all unusual to fall back on the hyperbolic vocabulary of "magic." The mind plays tricks on us. The way the brain produces consciousness is *quite* magical. Those who insist that consciousness is terminally mysterious, for instance, are wont to wallow in the stunning inexplicability of the effects known to us as the phenomenology of consciousness. And when one of these effects is explained, one can sometimes observe the same disappointment, the same resistance: to explain an effect is to diminish it.

Take déjà vu, for instance. Some incurable romantics have thought it a phenomenon at the magical end of the spectrum: according to them, we sometimes experience events that we know we have experienced before, in another life, in another astral plane, in another dimension. And they wonder what stunning insights this gives us into the cyclical nature of time, the transmigration of the soul, precognition, ESP . . . Pretty exciting stuff! But we can readily recognize that the phenomena of déjà vu could be explained in a much simpler way. You don't actually *remember* having experienced this very event as some time in the past; you just mistakenly *think* that you do. As Janet (1942) hypothesized more than half a century ago, it could be that it "results from an interruption of the perceptual process so that it splits into a past, as well as another current experience."

Figure 3.1 shows a simple diagram inspired by Janet's suggestion (for an earlier version, see Dennett 1979). Suppose that the visual system is redundant, containing two streams, A and B, which may be similar in their functions and powers or different, as you like. And suppose that both streams send their signals through a turnstile of sorts, a familiarity detector (or, alternatively, a novelty detector) that discriminates all incoming signals into those that are novel and those that have been encountered before. (There is evidence that triage of this sort

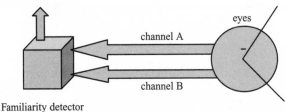

"I've seen it before!"

channel A

eyes

channel B

Familiarity detector
(hippocampus?)

Figure 3.1
Delayed channel model of déjà vu.

occurs quite early in visual processing, in the hippocampus, so this is not an entirely gratuitous speculation. See Gray 1995 and my 1995a commentary.) And let's suppose further that every now and then the transmission of signals through the B channel is ever so slightly delayed, so that it arrives at the familiarity detector a few milliseconds after the signal in the A channel. When the A channel signal arrives, it registers its novel footprint in the familiarity detector, and almost immediately that memory trace is discovered to match the signal now arriving on the B channel, triggering the familiarity detector to issue its positive verdict: "I've already seen this!" Not weeks ago, or months ago, or in a different life, but only a few milliseconds ago. What sequelae are provoked by this false alarm will depend on further details of the subject's psychology, ranging from slack-jawed wonder and exclamations about time travel to the slightest jaded smirk: "Oh, cool. I just had a déjà vu moment. I've seen *those* before, too!"

Such a simple transmission delay in a redundant system would be sufficient to explain the phenomenon of déjà vu, but if the two-channel model inspired by Janet's conjecture could explain it, so could the simpler, one-channel system shown in figure 3.2. In this simpler model, some perturbation or other—the death of a neuron, a neuromodulator imbalance, fatigue of one sort or another—could spuriously trigger a false positive verdict in the familiarity detector, and the rest of the sequelae could elaborate in whatever way they are supposed to elaborate in the other model. The main point to consider is that "from the inside," from the first-person point of view, the two models are indistinguishable. Nothing you can note about how déjà vu feels or seems to you could distinguish between the two models. If one of them (or some third or fourth model) is the truth, this

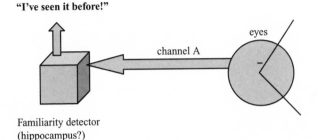

"I've seen it before!"

eyes

channel A

Familiarity detector
(hippocampus?)

Figure 3.2
Single channel model of déjà vu.

will have to be established by third-person investigations of the neural machinery in your head. We will have to go backstage to explain this particular bit of stage magic.

Another startling effect is the "filling in" we can think we discover in our own visual experience. The first time I spied Bellotto's *View of Dresden* (see figure 3.3) on a distant wall in the North Carolina Museum of Art in Raleigh, I took it for a Canaletto, and eagerly approached it, expecting to enjoy, up close, the exquisite detail that Canaletto lavished on his Venetian ships and gondolas, right down to the rigging lines, the buckles on the shoes, the plumes in the hats. The assorted crowd of people moving across the Dresden bridge in the sunlight promised a feast of costumes and carriages, but as I got closer, the details I could have sworn I had seen from afar evaporated before my eyes. Nothing but artfully placed simple blobs of paint were there to be seen up close (fig. 3.4).

Those spots "suggest" people, with arms and legs and clothes, and my brain had *taken the suggestion*. But what does that mean? What had my brain done? Sent out a team of homuncular

Figure 3.3
Bernardo Bellotto, *View of Dresden with Fraunkirche at Left,* 1747. Courtesy North Carolina Museum of Art. Purchased with funds from the state of North Carolina.

brain-artists to sketch in faces, hands and feet, hats and coats, in the appropriate parts of some retinotopic maps? This is an empirical question, and it is not one that I could answer from my putatively privileged perspective as the subject of this remarkable experience. (It is like the empirical question of whether the visual cortex creates rotating shapes when it succumbs to the illusion in figure 2.2 in the previous chapter.) Almost certainly, nothing of the kind happened in my brain. There is every reason to believe that no further *pictorial rendering* was done by the brain. When the brain *takes* the suggestion, the brain is forming a belief or expectation, not painting a picture for itself to look at. How do we know that such an expectation was created? It was exposed when it was violated, provoking a gasp of surprise from me; I had been expecting confirmation and elaboration of some speculative (involuntary,

Figure 3.4
Details from Bellotto, *View of Dresden.*

Figure 3.4
Continued.

unconscious) hypotheses about what I would soon see, and this
expectation was abruptly thwarted.[1]

1. Involuntary responses such as startle or laughter are excellent—and
underutilized—data in heterophenomenology. Such a response can
sometimes *convict* a subject of having a belief in spite of his adamant
denial, or just his inability to avow. An elderly member of my family,
in the late stages of Parkinson's disease, was rendered blind, almost
immobile and incapable of speech. There was often some doubt among
visitors about whether he was still conscious at all, but he was, right up
to the end, and this could be demonstrated quite conclusively by unob-
trusively sliding a joke—a pun or a double entendre—into the one-way
conversation and seeing the corners of his mouth turn up in an involuntary smile and a little crinkling at the edges of the eyes. I highly rec-
ommend the technique; it can bring home vividly that there is still
somebody at home in there, listening attentively to whatever you are
saying.

It wasn't *pure* hallucination; Bellotto did provide some dabs of paint for me to see, counting on my suggestibility to finish the job. The effect achieved is thus actually rather like one of the disqualified methods of performing the Indian Rope Trick: the posthypnotic suggestion, or the reporter taking a bribe—but it's not a complete fake since there was some visual presentation. But still, like a stage magician, the brain cheats! Many people, I have discovered, react to this suggestion with outraged disbelief: "Not *my* brain!" An understandable loyalty, but unwarranted and ungrounded. This is precisely what you *don't* know from personal ("first-person") experience. As Siegel says,

> Magic reveals how wrongly we remember what we have seen, discloses the way in which memory is the bearing and nursing mother of illusion. Memory is the magician's assistant, confederate, and shill. Hearing the description of a trick I've done, I'm amazed at what's described, at the way in which memory has tricked the spectator far more audaciously than I. (1991, p. 438)

2   Dismantling the Audience

There may well seem to be one residual—and glaring—problem with the heterophenomenological method: by taking the subject's word as constitutive, it seems to leave intact the one most problematic element of all—the audience watching the magic show. And as I have argued at length (1991, and many subsequent papers), this imagined showcase, the Cartesian Theater, where everything comes together for consciousness, must be dismantled. All the work done by the imagined homunculus in the Cartesian Theater must be distributed among various lesser agencies in the brain, *none of which is conscious*. Whenever that step is taken, however, the Subject vanishes, replaced by mindless bits of machinery unconsciously executing their tasks. Can this be the right direction for a theory of consciousness to take?

Here opinion is strikingly divided. On the one hand, there are those who join me in recognizing that *if you leave the Subject in your theory, you have not yet begun*! A good theory of consciousness *should* make a conscious mind look like an abandoned factory (recall Leibniz's mill), full of humming machinery and nobody home to supervise it, or enjoy it, or witness it.

Some people hate this idea. Jerry Fodor, for instance:

> If, in short, there is a community of computers living in my head, there had also better be somebody who is in charge; and, by God, it had better be me. (1998, p. 207)

As so often before, Fodor makes the valuable contribution here of exposing and endorsing the very idea that is causing all the trouble. He is far from alone in his anxiety about the loss of self portended by the dismantling of the Cartesian Theater, but he stands alone in his ability to articulate the misguided fear clearly and humorously. Robert Wright puts a different emphasis on much the same worry:

> Of course the problem here is with the claim that consciousness is "identical" to physical brain states. The more Dennett et al. try to explain to me what they mean by this, the more convinced I become that what they really mean is that consciousness doesn't exist. (2000, ch. 21, n. 14)

Recall Siegel's wry comment on "real" magic:

> "I'm writing a book on magic," I explain, and I'm asked, "Real magic?" By *real magic* people mean miracles, thaumaturgical acts, and supernatural powers. "No," I answer: "Conjuring tricks, not real magic." *Real magic*, in other words, refers to the magic that is not real, while the magic that is real, that can actually be done, is *not real magic*. (1991, p. 425)

Real consciousness, Wright cannot help but thinking, is something other than—and more marvelous than—physical brain states. "Stage" consciousness—the sort of consciousness that can be engineered out of the activities of brain machinery—isn't real consciousness. Jane Smiley, in an excerpt from her new book, *A Year at the Races* (Knopf, 2004), notes that some have doubted that horses are conscious, and then adds: "In fact, there are experts on human intelligence, like Daniel Dennett, who maintain that *humans* don't have consciousness either—that human consciousness is a false by-product of the workings of the brain" (Smiley 2004, p. 63). I don't maintain, of course, that human consciousness doesn't exist; I maintain that it is not what people often think it is. The insistence that consciousness *must* turn out to be something inexplicable, irreducible, transcendent sometimes rises to a fever pitch, as for instance in Voorhees:

> Daniel Dennett is the Devil. . . . There is no internal witness, no central recognizer of meaning, and no self other than an abstract "Center of Narrative Gravity" which is itself nothing but a convenient fiction. . . . For Dennett, it is not a case of the Emperor having no clothes. It is rather that the clothes have no Emperor. (2000, pp. 55–56)

But that is the beauty of it! In a proper theory of consciousness, the Emperor is not just deposed, but exposed, shown to be nothing other than a cunning conspiracy of lesser operatives whose activities jointly account for the "miraculous" powers of the Emperor. Banished along with the Emperor are what might be called the Imperial Properties: the two most mysterious varieties being the Qualia Enjoyed by the Emperor and the Imperial Edicts of Conscious Will.

For those who find this road to progress simply unacceptable, there is a convenient champion of the alternative option: *If you don't leave the Subject in your theory, you are evading the main issue*!

This is what David Chalmers (1996) calls the Hard Problem. He says that any theory that merely explains all the functional interdependencies, all the backstage machinery, all the wires and pulleys, the smoke and mirrors, has solved the "easy" problems of consciousness, but left untackled what he calls the Hard Problem.

3   The Tuned Deck

There is no way to nudge these two alternative positions closer to each other. No compromises are available. One side or the other is flat wrong. I have tried to show that however compelling the intuition in favor of Chalmers is, that intuition *must* be abandoned; the tempting idea that there is a Hard Problem is simply a mistake. I cannot prove this, and some who love the Hard Problem find my claim so incredible that they admit, with some hilarity, that they can't take it seriously. So I won't make the tactical error of trying to dislodge with rational argument a conviction that is beyond reason. That would be wasting everybody's time, apparently. Instead, I will offer up what I hope is a disturbing parallel from the world of card magic: The Tuned Deck.

> For many years, Mr. Ralph Hull, the famous card wizard from Crooksville, Ohio, has completely bewildered not only the general public, but also amateur conjurors, card connoisseurs and professional magicians with the series of card tricks which he is pleased to call "The Tuned Deck." . . . (Hilliard 1938, p. 517)

Ralph Hull's trick looks and sounds roughly like this:

> Boys, I have a new trick to show you. It's called "The Tuned Deck." This deck of cards is magically tuned [Hull holds the deck to his ear and riffles the cards, listening carefully to the buzz of the cards]. By

their finely tuned vibrations, I can *hear* and *feel* the location of any card. Pick a card, any card . . . [The deck is then fanned or otherwise offered for the audience, and a card is taken by a spectator, noted, and returned to the deck by one route or another.] Now I listen to the Tuned Deck, and what does it tell me? I hear the telltale vibrations, . . . [buzz, buzz, the cards are riffled by Hull's ear and various manipulations and rituals are enacted, after which, with a flourish, the spectator's card is presented.]

Hull would perform the trick over and over for the benefit of his select audience of fellow magicians, challenging them to figure it out. Nobody ever did. Magicians offered to buy the trick from him but he would not sell it. Late in his life he gave his account to his friend, Hilliard, who published the account in his privately printed book. Here is what Hull had to say about his trick:

> For years I have performed this effect and have shown it to magicians and amateurs by the hundred and, to the very best of my knowledge, not one of them ever figured out the secret. . . . *the boys have all looked for something too hard.* (Ibid., emphasis added)

Like much great magic, the trick is over before you even realize the trick has begun. The trick, in its entirety, is in the name of the trick, "The Tuned Deck," and more specifically, in one word—"The"! As soon as Hull had announced his new trick and given its name to his eager audience, the trick was over. Having set up his audience in this simple way, and having passed the time with some obviously phony and misdirecting patter about vibrations and buzz-buzz-buzz, Hull would do a relatively simple and familiar card presentation trick of type A. (At this point I will draw the traditional curtain of secrecy; the further mechanical details of legerdemain, which are available to the curious in any good book on card magic, do not matter, as you will see). His audience, savvy magicians, would see that he might possibly be performing a type A trick, a hypothesis that they could test by being stubborn and uncooperative spectators in a way that would thwart any attempt at a type A trick. When they then adopted the appropriate recalcitrance to test the hypothesis, Hull would "repeat" the trick, this time executing a type B card presentation trick. The spectators would then huddle and compare notes: we've proved he's not doing a type A trick. Might he be doing a type B trick? They test *that* hypothesis by adopting the recalcitrance appropriate to preventing a type B trick and still he does "the" trick—using method C, of course. When they test the hypothesis that he's pulling a type C trick on them, he switches to method D—or perhaps he goes back to method A or B, since his audience has "refuted" the hypothesis that he's using method A or B. And so it would go, for dozens of repetitions, with Hull staying one step ahead of his hypothesis-testers, exploiting his realization that he could always do *some trick or other* from the pool of tricks they all knew, and concealing the fact that he was doing a grab bag of different tricks by the simple expedient of the definite article: *The* Tuned Deck.

> . . . each time it is performed, the routine is such that one or more ideas in the back of the spectator's head is exploded, and sooner or later he will invariably give up any further attempt to solve the mystery. (Ibid., p. 518)

I am suggesting, then, that David Chalmers has (unintentionally) perpetrated the same feat of conceptual sleight-of-hand in declaring to the world that he has discovered "The Hard Problem." Is there *really* a Hard Problem? Or is what appears to be *the* Hard Problem simply the large bag of tricks that constitute what Chalmers calls the Easy Problems of Consciousness? These all have mundane explanations, requiring no revolutions in physics, no emergent novelties. They succumb, with much effort, to the standard methods of cognitive science. I cannot prove that there is no Hard Problem, and Chalmers cannot prove that there is. He can appeal to your intuitions, but this is not a sound basis on which to found a science of consciousness. We have seen in the past—and I have given a few simple examples here—that we have a powerful tendency to inflate our inventory of "known effects" of consciousness, so we must be alert to the possibility that we are being victimized by an error of arithmetic, in effect, when we take ourselves to have added up all the Easy Problems and discovered a residue unaccounted for. That residue may already have been accommodated, without our realizing it, in the set of mundane phenomena for which we already have explanations—or at least, will be accommodated in unmysterious paths of explanation still to be explored.

The "magic" of consciousness, like stage magic, defies explanation only so long as we take it at face value. Once we appreciate all the nonmysterious ways in which the brain can create benign "user-illusions" (Dennett 1991, pp. 309–314), we can begin to imagine how the brain creates consciousness. But still, I know many are asking themselves, doesn't this reply obviously—obviously!—leave out the Master Illusion itself, the consciousness *I* know from the inside, the consciousness Descartes made famous with his meditation on *cogito ergo sum*? This is the Zombic Hunch, doggedly returning, and living off our habitual inability to keep track of all the cards on the table. This "I" you speak of is not some pearly something outside the physical world or something in addition to the team of busy, unconscious robots whose activities compose *you*, and hence it should not be left out of the accounting.

1 The Quale, An Elusive Quarry

The term "residue" often comes up in the philosophical literature on qualia, a mark of the popularity of Sherlock Holmes's
tactic of using a process of elimination to zero in on the elusive
quarry. And almost as frequent are the metaphors suggesting
that qualia are a sort of stuff, perhaps a liquid (if not ecto-
plasm!). The imagery is not restricted to philosophers; Rodney
Brooks sometimes expresses his wonder about whether robotics
and AI can deliver what he likes to call "the Juice" of consciousness. Gabriel Love has coined an acronym, *SAUCE*, which
stands for Subjective Aspect Unique to Conscious Experience, to
refer to this curious way of posing the issue. Let's go hunting
for the secret sauce.

As a left-handed person, I can wonder whether I am a left-
hemisphere-dominant speaker or a right-hemisphere-dominant
speaker or something mixed, and the only way I can learn the
truth is by submitting myself to objective, "third-person"
testing. I don't "have access to" this intimate fact about how
my own mind does its work. It escapes all my attempts at introspective detection, and might, for all I know, shunt back and

forth every few seconds without my being any the wiser. There is nothing unusual about this fact; most of the events occurring in my body, and indeed in my brain, occur without my knowledge. In striking contrast to this unsurprising state of affairs, there are, however, some events that occur in my brain that I do know about, as soon as they occur: my subjective experiences themselves. And these subjective experiences, tradition tells us, have "intrinsic qualities"—qualia, in the jargon of philosophers—that I not only do have access to, but that are inaccessible to *objective* investigation. This idea has persisted for centuries, in spite of its incoherence, but perhaps its days are finally numbered.

Are there *qualia*? If so, just what are they? "Qualia" is the plural of "quale," a Latin word for *quality*, but philosophers have endowed the term with a variety of ill-considered associations and special powers. Since the term has no home use other than in the philosophy of mind, we really have no choice other than to let philosophers define it as they will, if only they would! But there is no agreed-on definition. This does not bother philosophers as much as it might be expected to. In his essay on qualia in the *Stanford Encyclopedia of Philosophy* (available online at http://plato.stanford.edu/entries/qualia), Michael Tye introduces the concept as follows:

> Feelings and experiences vary widely. For example, I run my fingers over sandpaper, smell a skunk, feel a sharp pain in my finger, seem to see bright purple, become extremely angry. In each of these cases, I am the subject of a mental state with a very distinctive subjective character. There is something it is like for me to undergo each state, some phenomenology that it has. Philosophers often use the term "qualia" (singular "quale") to refer to the introspectively accessible, phenomenal aspects of our mental lives. In this standard, broad sense of the term, it is difficult to deny that there are qualia.

Yes, it is indeed difficult to deny that there are qualia. I've been working on the task for years, with scant progress! The reason it is difficult is mainly that this "standard, broad sense of the term" is a conspiracy of unexamined presuppositions and circularly defined elaborations. Just how "introspectively accessible" must an aspect be to count as a quale? Which aspects of our experiences are the "phenomenal aspects" and which are not? Is our *enjoyment* of a good meal, for instance, *itself* a phenomenal aspect of the experience, or is it an *effect* of, or a *response* to, a phenomenal aspect (the deliciousness, let's say)? If the enjoyment were somehow obtunded, would the deliciousness still be present but just sadly unappreciated? (Can there be unfelt pains, and if so, do they have—or are they— qualia, or are these pains only the normal causes of qualia?) What does "phenomenal" mean? "Phenomenal" aspects or properties are usually contrasted with "relational" or "functional" properties of experience, but this negative definition is unsatisfactory—as uninformative as the claim that the "spiritual" properties of a person are those that are *not* physical. (Your body is composed of roughly a hundred trillion cells. As cells die and sometimes get replaced, this number no doubt fluctuates, changing every millisecond. Every now and then, there will be a brief period of time when the number of cells in your body will be a prime number. Is this one of your spiritual properties? If not, why not?)

Until "phenomenal" is positively defined, we can't really evaluate the claims made about phenomenal aspects, and if the term is interdefined with "qualia" then we are still in the dark about just what we are talking about when the topic is qualia. It also does not help us when we are told that qualia are what zombies don't have, though this seems to be growing in popularity as a way of *pointing to* the elusive quarry, while awaiting a satisfactory definition. Nobody in philosophy thinks that there are actually any zombies, but many philosophers think it is important to consider the (logical) possibility that there *could be* zombies, and what the implications of this possibility are. And one of the most popular purported implications is that the possibility of zombies exposes a fatal defect in heterophenomenology as a method for studying consciousness.

Everybody's favorite examples of qualia are "subjective colors," such as the luscious yellow you enjoy when you look at a ripe lemon or the breathtaking shade of warm pink you see in the western sky during a glorious sunset. The color qualia are not the objective features of the light, the features captured on color film or color videotape; they are supposedly the purely subjective effects in you of seeing the lemon, or the photograph of the lemon, or the videotape of the lemon. But just moving the qualia inside the mind in this way doesn't come close to settling just which sort of effects the qualia are meant to be. They are *in you* and hence potentially *idiosyncratic* but presumably not all idiosyncratic properties produced in you by your sensory systems are qualia. As Stephen Palmer (1999) puts it, in an important essay in *Behavioral and Brain Sciences*, the traditional view is that "I alone have access to these experiences" (p. 938). But this obvious-sounding claim must be defended against the apparently unthinkable hypothesis that *not even I* "have access to" the intrinsic qualities of my very own experience. What could this mean? It could mean that there were intrinsic qualities of my experience whose comings and goings were, like the spatial properties of my language-comprehension and production activities, beyond my direct ken. This invites the obvious retort: then they wouldn't be properties *of my experience*! And what could that mean?

Palmer focuses his attention on within-subject experiments, as contrasted with between-subject experiments—which raise notorious problems of intersubjective comparison—and notes that even in a within-subject experiment, the individual subject must make a "memory comparison," and Palmer acknowledges the theoretical possibility that there might be intrinsic qualities that changed so gradually, over such a long time, that the intra-subjective memory comparison would fail to detect them. If a change were slow enough, he concedes, even a huge change could occur without being detected, and if a change were subtle enough, it could change quickly without the subject's noticing. But never mind, he says, since he is concerned only with those within-subject changes in experiential quality that are "swift and enormous" (p. 939). How swift and how enormous? Just swift and enormous enough *to be detected by the subject.*

Palmer concludes that "[w]ithin-subject designs can examine changes in experience, but cannot reveal what they changed from or to" (p. 942). But then notice that you are in the same predicament as the experimenter; you do *not* "have access to" the intrinsic qualities of your own experiences in any interesting sense, any more than outside observers do. You have access only to the relations between them *that you can detect.* The very detectability *by the subject* of "swift and enormous" changes guarantees that any such changes of properties are "within the domain of functionalism"—they are objectively investigatable by standard heterophenomenological methods.[1]

1. Material in the previous three paragraphs is drawn, with revision, from Dennett 1999.

This point does not establish that there are no intrinsic qualities of experience that are *not* accessible to heterophenomenology—Palmer calls such properties "subisomorphic"—but only that if there are, their presence or absence is something to be determined indirectly by third-person scientific investigation and theory, since they make no difference *discernible by the subject* to the subjective state of the subject. There are plenty of subisomorphic properties of experiences that are readily detectable by *objective* probes of one sort or another (such as the chemical constitution of the neuromodulator molecules involved), but are these "intrinsic" properties of the experience in the sense intended? I doubt it. (We are all free to believe that our experiences have "intrinsic" properties that are unknown to us—just as we are free to believe that there are planets outside our light cone inhabited by talking rabbits—but these would be facts that could not make any difference to us.)

Even "swift and enormous" changes can escape the detection of subjects, however, as has been shown in the phenomenon of *change blindness*, predicted by me in *Consciousness Explained* (p. 468) and subsequently explored in dozens or perhaps hundreds of experiments by Grimes, Rensink, O'Regan, Simons, and many others.

2 Change Blindness and a Question about Qualia

In recent years I have often shown philosophical audiences a videotape of an early experiment by Rensink, O'Regan, and Clark (1997) in which pairs of near-twin photographs are shown for 250 milliseconds each, separated by a similarly brief (290 milliseconds) blank screen (a "mask") and repeated in alternation until the subject sees the change between the A picture and the B picture and presses a button. Subjects often study these alternating pictures for twenty or thirty seconds, with the change displayed dozens of times before their very eyes, before being able to spot the change. In my presentation, members of the audience can play the role of informal subject. And one of the pairs of pictures in the videotape I have shown has a particularly hard-to-spot change of color. It is a photograph of a kitchen, and one of the cabinet doors flashes back and forth between white and brown (as in figure 4.1). It is not a small change, and once you notice it, it is hard to believe it has gone unmarked in your experience for some dozens of repetitions.

Having exposed the audience to the experience (and having finally drawn their attention to the flashing door, since thirty seconds is a long time to wait for them to tumble!), I ask them all a question:

> Now *before you noticed* the panel changing color, were your color qualia for that region changing? We know that the cones in your retinas in the regions where the light from the panel fell were responding differently every quarter of a second, and we can be sure that these differences in transducer output were creating differences farther up the pathways of color vision in your cortex. But were your *qualia* changing back and forth—white/brown/white/brown—in time with the color changes on the screen? Since one of the defining properties of qualia is their subjectivity, their "first-person accessibility," presumably nobody knows—or *could* know—the answer to this question better than you. So what is your answer? Were your qualia changing or not?

This is an atypical question for a heterophenomenologist to ask, since it invokes "qualia," a theoretical term of uncertain standing which invites theorizing, not unvarnished phenomenological reporting, by the subjects. (There's no harm in asking it, but one should be wary of the answers, since they will be

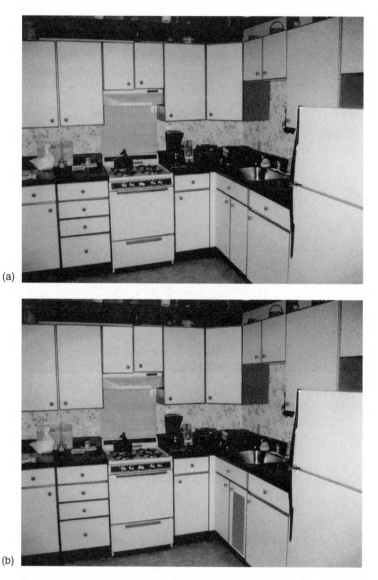

(a)

(b)

Figure 4.1a, b
A pair of change blindness pictures similar to those shown in the demonstration.

hard to disambiguate, as we shall see.) There are three possible answers:

A. Yes.

B. No.

C. I don't know,

(1) because I now realize I never knew quite what I meant by "qualia" all along;

(2) because although I know just what I have always meant by "qualia," I have no first-person access to my own qualia in this case;

(a) and third-person science can't get access to qualia either!

Put yourself in the subject's position and consider how you would answer. (If you haven't yet had the experience of change blindness yourself, you ought to be able to imagine just *what it's like* from the description I've given: until you notice the change, it's indistinguishable from looking at an unchanging picture flashing in alternation with a mask, and surely you know what that's like). All three answers have their problems. If you are inclined to answer *Yes*, you are constrained to admit that swift and enormous changes in your qualia can occur without your knowledge. You must countenance the possibility that you often, even typically, are oblivious to large sudden shifts in your qualia. This would undermine the standard presumption that you are authoritative or even incorrigible about them. Others, *third persons* or Martians, even, might be better authorities than you are about the constancy or inconstancy of your own qualia.

So perhaps you are therefore inclined to answer *No*. Then you can retain your authority about your qualia by claiming that *since* you noticed no shift, your qualia didn't shift, no matter what else in your brain shifted. Since you are the subject, your subjective state—colors-apparently-shifting or colors-apparently-constant—can be held to be definitive of your color qualia. This claim, however, threatens to trivialize qualia as just logically *constituted* by your judgments or noticings, an abandonment of the other canonical requirement for qualia: that they be "intrinsic" properties. More pointedly, if you maintain that your qualia shift if and only if you think they do, and stay constant otherwise, you will also have to abandon the idea that zombies lack qualia. A zombie would be just as subject to change blindness as any normally conscious being, because zombies are behaviorally indistinguishable from normal human beings. Putting subjects in a change blindness experiment will not provide the slightest leverage in sorting the sheep from the goats. A zombie *thinks* it has qualia and either thinks they are shifting or doesn't. Why would a zombie's judgments be any less authoritative than yours? (And if zombies are not authoritative about their qualia judgments, how do you know you're not a zombie?)

Let's turn to option C then. If, confronted with this problem, you decide that *you don't know* whether your qualia were shifting before you noticed the change, you put qualia in the curious position of being beyond the horizon of both third-person objective science and first-person subjective experience. I have found, in fact, that people confronted with these three choices don't agree; all three answers find supporters who are, moreover, typically surprised to find that the other two answers have any takers at all. This informal finding supports my long-standing claim (Dennett 1988) that philosophers actually don't know what they are talking about when they talk about their qualia.

Many people discover, when they confront this case, that since they never imagined such a phenomenon was possible, they never considered how their use of the term "qualia" should describe it. They discover a heretofore unimagined flaw in their concept of qualia—rather like the flaw that physicists discovered in their concept of *weight* when they first distinguished weight from mass.

The philosophers' concept of qualia is a mess. Philosophers don't even agree on how to apply it in dramatic cases like this. This should be at least mildly embarrassing to our field, since so many scientists have recently been persuaded by philosophers that they should take qualia seriously—only to discover that philosophers don't come close to agreeing among themselves about *when* qualia—*whatever* they are—are present. I have found that many scientists who think they are newfound friends of qualia turn out to use the term in ways no self-respecting qualophile philosopher would countenance. Palmer is a case in point. In his (1999) reply to Dennett 1999, he misinterprets my claim that we have access only to the relational properties of our experiences. He finds it "possibly incoherent" because he thinks I am speaking of such relational properties as those explored by Land in his retinex theory of color: the "contrast ratios between the luminances of adjacent regions" (p. 978) in a visible scene, for instance. I am referring rather to such relational properties as those that hold between sensory states and the beliefs they normally cause in subjects, the functionalism-friendly properties that can be designed into information-processing systems. When Palmer says "The lightest rectangle in an achromatic Mondrian looks *white*, not just twice as light as the next-darker rectangle" (p. 978), he is right, but only in a sense that is not in fact a rebuttal of my claim. He is supposing that *looking white* is obviously an intrinsic property, but that is just the supposition I am calling into question. One could interpret *looking white* as an intrinsic property, no doubt, but then it would have to be sharply distinguished from the relational property of *looking white to Jones at time t*, which is the property philosophers are interested in. The change blindness example can be used to bring this out quite sharply. Does the kitchen cabinet door cease to *look white* twenty times in less than twenty seconds? Yes, on the screen. Does it cease to look white to Jones? Once Jones notices the change, it ceases to look white to Jones on many subsequent occasions; but what should Jones say about this property of *looking white to him* (the subjective property) before he noticed the change?

Although some philosophers may now concede that they aren't so sure what they meant by "qualia" all along, others will claim to be very sure what concept of qualia they've been using all along, so let's consider what they say. Some of them, I have learned, have *no problem* with the idea that their very own qualia could change radically without their noticing. *They* mean by "qualia" something to which their first-person access is variable and problematic. If you are one of those, then heterophenomenology is your preferred method, since it, unlike the first-person point of view, can actually study the question of whether qualia change in this situation. It is going to be a matter of some delicacy, however, how to *decide* which brain events count for what. In this phenomenon of change blindness for color changes, for instance, we know that the color-sensitive cones in the relevant region of your retina were flashing back and forth, in perfect synchrony with the white/brown quadrangle, and presumably (we should check) other, later areas of your color vision system were also shifting in time with the external color shift. But if we keep looking, we will also presumably find yet other areas of the visual system that only come into synchrony after you've noticed (such effects have been found in similar fMRI studies, e.g., O'Craven et al. 1997).

The hard part will be deciding (on what grounds?) which features of which states to declare to be qualia and why. I am not saying there can't be grounds for this. I can readily imagine there being *good* grounds, but if so, then those will be grounds for adopting and endorsing a third-person concept of qualia (see, e.g., the discussion of Chase and Sanborn in Dennett 1988, or the example of the beer drinkers in Dennett 1991, pp. 395–396). The price you have to pay for obtaining the *support* of third-person science for your conviction about how it is/was with you is straightforward: you have to grant that *what you mean* by how it is/was with you is something that third-person science could either support or show to be mistaken. Once we adopt any such concept of qualia, for instance, we will be in a position to answer the question of whether color qualia shift during change blindness. And if some subjects in our apparatus tell us that their qualia do shift, while our brainscanner data show clearly that they don't, we'll treat these subjects as simply wrong *about their own qualia*, and we'll have to explain why and how they come to have this false belief.

Some people find this prospect inconceivable. For just this reason, these people may want to settle for option B: *No*, my qualia don't change—*couldn't* change—until I notice the change. This decision *guarantees* that qualia, tied thus to noticing, are securely within the heterophenomenological worlds of subjects, are indeed *constitutive* features of their heterophenomenological worlds. On option B, what subjects can say about their qualia fixes the data.[2]

If, then, this continuing cycle of the process of elimination brings us back again to option A, *Yes*, the heterophenomenologist will want to ask you some further questions: if you think your qualia did change (though you didn't notice it at the time) *why* do you think this? Is this a theory of yours? If so, it needs evaluation like any other theory. If not, did it just come to you? A gut intuition? Either way, your conviction is a prime candidate for heterophenomenological diagnosis: what has to be explained is how you came to have this belief. The last thing we want to do is to treat your claim as incorrigible. Right?

Here is the dilemma for those who are attracted to a first-person view. If you eschew incorrigibility claims, and especially if you acknowledge the competence of third-person science to answer questions that can't be answered from the first-person point of view, your position collapses into heterophenomenology. The only remaining alternative, C(2a), is unattractive for a

2. Consider option B for the simpler case raised earlier (in chapter 2, p. 41). Do you want to cling to a concept of visual consciousness according to which your conviction that your visual consciousness is detailed all the way out is *not* contradicted by the discovery that you cannot identify large objects in the peripheral field? You *could* hang tough: "Oh, all that you've shown is that we're not very good at identifying objects in our peripheral vision; *that* doesn't show that peripheral consciousness isn't as detailed as it seems to be! All you've shown is that a *mere behavioral capacity* that one might mistakenly have thought to coincide with consciousness doesn't, in fact, show us anything about consciousness!" Yes, if you are careful to *define* consciousness so that nothing "behavioral" can bear on it, you get to declare that consciousness transcends "behaviorism" without fear of contradiction. See chapter 7 for a more detailed account of this occasionally popular but hopeless move.

different reason. You can protect qualia from heterophenome-nological appropriation, but only at the cost of declaring them outside science altogether. If qualia are so shy they are not even accessible from the first-person point of view, then no *first-person* science of qualia is possible either.

I will not contest the existence of first-person *facts* that are unstudiable by heterophenomenology and other third-person approaches. As my colleague Stephen White has reminded me, these would be like the humdrum "inert historical facts" I have spoken of elsewhere (e.g., Dennett 2003)—like the fact that some of the gold in my teeth once belonged to Julius Caesar, or the fact that none of it did. One of those is a fact, and I daresay no possible extension of science will ever be able to say which is the truth. But if first-person facts are like inert historical facts, they are no challenge to the claim that heterophenomenology is the maximally inclusive science of consciousness, because they are unknowable even to the first person they are about!

3   Sweet Dreams and the Nightmare of Mr. Clapgras

One of the themes about qualia that is often presupposed but seldom carefully discussed was memorably made explicit for me by Wilfrid Sellars, over a fine bottle of Chambertin, in Cincinnati in 1971: I had expressed to him my continuing skepticism about the utility of the concept of qualia and he replied: "But Dan, qualia are what make life worth living!" (Dennett 1991, p. 383).

The basic idea is clear enough. If you didn't have qualia, you would have nothing to *enjoy* (but also no suffering, presumably). It is generally supposed—though seldom if ever expressed—that it would not be any fun to be a zombie. Nobody wants to become a zombie. Being a zombie would be like being a telephone pole—like nothing at all. It isn't like anything to be a zombie, so of course it isn't fun being a zombie. But at least a zombie wouldn't suffer. If qualia are what make life worth living, then zombies' lives are not worth living. You get the idea.

Except you *don't* get the idea. Or at least I doubt that you do; I doubt that anybody who *gets* the idea of a zombie, an agent without qualia, in its full implications, can fail to recognize that it is an irreparably incoherent idea. To bring out the covert contradictions in the very idea of a zombie—and hence the very idea of qualia in at least one of its most popular senses—I want to explore rather more directly what would have to be the case if, as Sellars said, qualia were what made life worth living. To see what is at issue, I will present a new thought experiment against a background of recent work in cognitive neuroscience on several bizarre and counterintuitive pathologies: *prosopagnosia* and *Capgras delusion*.

Prosopagnosics have normal vision with one strange disability: they cannot recognize faces. They can tell a male from a female, old from young, African from Asian, but faced with several close friends of the same gender and age, they will be unable to tell which is which—until they hear a voice or detect some other identifying peculiarity. Given a row of photographs of people, including famous politicians and movie stars, family members, and anonymous strangers, a prosopagnosic asked to identify any that are known to him will generally perform at chance. Those of us who are not prosopagnosics may find it difficult to imagine what it can possibly be like to look right at one's mother, say, and be unable to recognize her. Some may then find it hard to believe that there could be such a complaint as prosopagnosia. When I tell people about these phenomena,

I often discover skeptics who are quite confident that I am simply making these facts up! But we must learn to treat such difficulties as measures of our frail powers of imagination, not insights into impossibility. Prosopagnosia is a well-studied, uncontroversial pathology afflicting thousands of people.

One of the most interesting facts about some prosopagnosics is that in spite of their inability to identify or recognize faces as a conscious task, they can be shown to respond differently to familiar and unfamiliar faces, and even to respond in ways that show that *unbeknownst to themselves*, or *covertly*, they were correctly identifying the faces that they were unable to identify if asked. For instance, such "covert recognition" is demonstrated when prosopagnosics are shown pictures and given five candidate names from which to choose. They choose at chance, but their galvanic skin response—a measure of emotional arousal—shows a distinct rise when they hear the correct name. Or consider this simple test: Which of the following are *names* of politicians: Marilyn Monroe, Al Gore, Margaret Thatcher, Mike Tyson? An easy task, which you can answer swiftly, but your response is markedly delayed on a name if the wrong picture is associated with it. This could be explained only if *at some level*, the subjects were actually identifying the faces. It seems, then, that there are (at least) two largely independent visual face-recognition systems in the brain: the impaired "conscious" system, which cannot help the subjects in the task set them by the experiment, and the unimpaired "unconscious" system, which responds with agitation to the mismatched names and faces. Further tests show that the impaired system is "higher"—in the visual cortex—while the unimpaired system has connections to the "lower" limbic system. This oversimplifies a richer story about the varieties of prosopagnosia and what is now known about the brain areas involved, but it will do for our purposes, as we turn to the even stranger pathology known as Capgras delusion.

People who suffer from Capgras delusion suddenly come to believe that a loved one—a spouse or lover or parent, in most cases—has been covertly replaced with a replica impostor! Capgras sufferers are not hysterical or insane; they are otherwise *quite* normal people, who, as a result of brain injury, suddenly acquire this particular belief, which they maintain with such confidence, in spite of its extravagance and its utter unlikeliness, that there have been cases in which the "impostor" has been killed or seriously harmed by the deluded sufferer. At first glance it must seem simply impossible for any brain damage to have precisely *this* weird effect. (Should we also expect there to be people who get hit on the head and thereafter believe that the moon is made of green cheese?) But Andrew Young saw a pattern, and proposed that the Capgras delusion was basically the "opposite" of the pathology that produces prosopagnosia. In Capgras, the conscious, cortical face-recognition system is spared—that's how the person recognizes the person standing in front of him as the spitting image of his loved one—but the unconscious, limbic system is disabled, draining the recognition of all the emotional resonance it ought to have. The *absence* of that subtle contribution to identification is so upsetting ("Something's missing!") that it amounts to a pocket veto on the positive vote of the surviving system: the emergent result is the sufferer's heartfelt conviction that he or she is looking at an impostor. Instead of blaming the mismatch on a faulty perceptual system, the agent blames the world, in a way that is so metaphysically extravagant, so improbable, that there can be little doubt of the power (the political power, in effect) that the impaired system normally has in us all. When this particular system's epistemic hunger goes unsatisfied, it throws such a fit that it overthrows the contributions of the other systems.

Since Ellis and Young first proposed this hypothesis in 1990, it has been elaborated and confirmed by Young and others (see, e.g., Burgess et al. 1996; Ellis and Lewis 2001). There are, of course, complications that I will not dwell on, since I want to use this particular bit of imagination-stretching cognitive neuroscience to open our minds to yet another possibility, not yet found but imaginable. This is the *imaginary* case of poor Mr. *Clapgras*, a name I have made up to remind us of its inspiration: the real syndrome of Capgras delusion.

Mr. Clapgras earns a modest living as an experimental subject in psychological and psychophysical experiments, so he is far from naive about his own subjective states. One day he wakes up and cries out in despair as soon as he opens his eyes: "Aargh! There's something wrong! The whole world is just. . . . *weird*, just . . . *awful*, somehow *wrong*! I don't know if I want to go on living in *this* world!" Clapgras closes his eyes and rubs them; he cautiously opens them again, only to be confronted yet again by a strangely disgusting world, familiar but also different in some way that defies description. That is what he says, and his heterophenomenological interlocutors are frankly puzzled. "What do you see when you look up?" he is asked. "Blue sky, with a few fleecy white clouds, some yellowish-green buds on the springtime trees, a bright red cardinal perched on a twig," he replies. Apparently his color vision is normal, but just to check, he is given the standard Ishihara test, which shows he is not color-blind, and he correctly identifies a few dozen Munsell color chips. Almost everybody is satisfied that whatever poor

Mr. Clapgras's ailment is, it doesn't involve his color vision, but one researcher, Dr. Chromaphil, holds out for a few more tests.

Dr. Chromaphil has been conducting research on color preferences, emotional responses to color, the effects of different colors on attention, concentration, blood pressure, pulse rate, metabolic activity, and a host of other subtle visceral effects. Over the past six months he has accumulated a huge database about Mr. Clapgras's responses, both idiosyncratic and common, on all these tests, and he wants to see if there have been any changes. He retests Clapgras and notices a stunning pattern: all the emotional and visceral responses Clapgras used to exhibit to blue he now exhibits to yellow, and vice versa. His preference for red over green has been reversed, as have all his other color preferences. Food disgusts him—unless he eats in the dark. Color combinations he used to rate as pleasing he now rates as jarring—while finding the combinations of their "opposites" pleasing, and so forth. The shade of shocking pink that used to set his pulse racing he still identifies as shocking pink (though now he marvels that anybody could call *that* shade of pink shocking), while its complement, a shade of lime green that used to be calming to him, is now exciting. When he looks at paintings, his trajectory of saccades is now profoundly unlike his earlier trajectories, which were apparently governed by subtle attention-grabbing, gaze-deflecting effects of the colors on the canvas. His ability to concentrate on mental arithmetic problems, heretofore seriously depressed by putting him in a bright blue room, is now depressed by putting him in a bright yellow room.

In short, although Clapgras does not complain about any problems of color vision, and indeed passes all standard color-naming and color-discriminating tests with, well, flying colors, he has undergone a profound inversion of all his emotional and attentional reactions to colors. What has happened to Clapgras, according to Dr. Chromaphil, is simple: he's undergone a total color *qualia* inversion, while leaving his merely high-level cognitive color talents—his ability to discriminate and name colors, for instance—intact.

Digression: In chapter 2 I drew attention to the task we theorists all face of accomplishing a division of labor that permits all the work to be done by the imaginary homunculus or Central Witness in the Cartesian Theater to be broken into subtasks and outsourced, as the businesspeople say these days: distributed to lesser specialists in the brain. Chromaphil's catalog of effects provides a partial list of some of the *color-appreciation-and-discrimination* jobs traditionally assigned to the Central Witness, and we are supposing that some of them might be dissociated from others and then inverted relative to their previous outcomes. That some of these might be spatially isolated in the brain is not implausible, given what we are learning about parallel streams of functional specialists working on other tasks, such as navigation, danger-alerting, and face-recognition.

Now what should we say? Have Clapgras's qualia been inverted? Since the case is imaginary, it seems that we may answer it however we like, but philosophers have been taking other imaginary cases seriously for years, thinking that profound theoretical issues hinge on how they are decided, so we mustn't just dismiss the case as a fantasy. First, is this a *possible* case? It may depend on what kind of possibility we are talking about. Is it logically possible? Is it physiologically possible? These are profoundly different questions. Philosophers have tended to ignore the latter sort as quite irrelevant to philosophical concerns, but in this case, they may relent. I can see no way of arguing that the case is logically impossible. Clapgras, as described, has a strange combination of spared abilities and shocking new inabilities; dispositions that are normally tightly linked are here dissociated in unprecedented ways; but is his condition any more radical in this regard than either prosopagnosia or Capgras delusion? I am not sure Clapgras's condition is even physiologically impossible; there are well-studied cases of subjects who can discriminate colors just fine but cannot name them (color anomia), and of subjects who become color-blind but don't notice their new deficit, blithely confabulating and naming colors at chance without any recognition that they are guessing. Clapgras, like a Capgras sufferer, has no problems with recognition or naming; it is the subtle ineffable flavoring that has gone all askew in him—all the personal dispositions that make paintings worth looking at, rooms worth painting, color combinations worth choosing. The subjective effects of colors that contribute to making life worth living have changed in Clapgras—in other words (if Sellars is right), his color *qualia*.

But as before, in the case of change blindness, we should put the issue to Clapgras and ask him if his color qualia have been inverted. He has three possible answers: *Yes*, *No*, and *I don't know*. Which should he answer? If we compare my story of Clapgras with the many tales of inverted qualia that have been carefully promulgated and discussed at great length by philosophers, the most disturbing innovation is the prospect that Clapgras might have his qualia inverted and be none the wiser. Dr. Chromaphil has to propose this hypothesis to his skeptical colleagues, and Clapgras may well share their skepticism. After all, he not only hasn't complained of any problem with his color qualia (as in the standard stories), he in fact satisfies himself that his color vision is just fine in the same way he satisfies the researchers: by easily passing the standard color vision tests. This ought to cause some discomfort in philosophers: surely *those* tests have no bearing at all on qualia, or at least so it is commonly assumed in the philosophical literature. Those tests are standardly characterized as having no power to illuminate or constrain qualia quandaries. But, as my variation shows, philosophers' imaginations have overlooked the prospect of somebody's being at least *tempted* to rely on these tests to secure his own confidence that his qualia have not changed. Is someone thus tempted simply confused? If he shouldn't rely on his willingness and ability to name colors just as he used to name them, what should he rely on? Can he *just tell* that his memory of what yellow used to look like to him is, or is not, what it looks like to him now? I gather from discussions of this thought experiment with philosophers that there is an awkward temptation here to suppose that you can just set yourself the task of *imagining yellow* (say) and *know* that you're doing the very same thing you've always done when you imagined yellow in the past. But if you find that you have this intuition, you ought to discard it immediately, or at least reserve your allegiance to it.

Once again, I find that philosophers divide over how to answer certain definitional questions about qualia, in particular:

Can your qualia stay constant while you undergo a change in "affect"?

Consider the effect of monosodium glutamate, the flavor enhancer. There is no doubt that it makes food seem tastier, more strongly flavored, but which of several apparently different imaginable phenomenological effects does it have? Does it change the *qualia* of food (the way table salt does, by adding the *salty* quale, or the way sugar does by adding the *sweet* quale—I'm presuming that this is how qualophiles would put it), or does it merely heighten the sensitivity of people to the qualia they were already enjoying? Does it add a new "intrinsic property" or does it just help the subject make contact with the intrinsic properties that are somehow already *there* in consciousness? Recent research reveals that there are specific receptor proteins for glutamate on mammalian tongues, similar to those for detecting sweetness, the *savory* or *umami* detectors (Kawamura and Kare 1987; Rolls and Yamamoto 2001; Zhao et al. 2003). Is there an umami quale, then? The downstream effects of umami-receptor excitation are beginning to be mapped, and sure enough, they turn on a preference for umami-laced food over plainer fare, but this "merely behavioral" evidence doesn't—couldn't, according to the qualophiles—settle the issue. What is needed here is not (just) more empirical research on the sites of action of these downstream effects, but a clarification of the concept of qualia. This will have to be legislated, since there is no consensus among philosophers about how to use the term: should we identify all changes in subjective response as changes in qualia, or is there some privileged subset of changes that anchor the qualia? Is the idea of changing one's aesthetic opinion about—or response to—a particular (constant) quale nonsense or not? Until one makes decisions about such questions of definition, the term is not just vague or fuzzy; it is radically ambiguous, equivocating between two (or more) fundamentally different ideas. There is no point in continuing to use the term until these equivocations are cleared up, one way or another.

Back to poor Mr. Clapgras. I find that when I ask philosophers which way they would jump in describing his predicament, some object that I haven't given enough detail in describing his condition. I have described his behavioral competences—he recognizes, and discriminates, and names colors correctly, while responding "anomalously" in many other regards—while deliberately avoiding describing his subjective state. I haven't said whether or not, for example, when he looks at a ripe lemon, he experiences *intrinsic subjective yellow* or, say, *intrinsic subjective blue*. But that is the point: I am challenging the presumption that these terms name any real properties of his experience at all. If outsiders cannot tell, no matter how fine-grained their maps of the functional neuroanatomy, and if Clapgras himself cannot tell, then perhaps these intrinsic properties are an artifact of an obsolete theoretical vision—the Cartesian Theater—rather than anything that we should continue to seek in our scientific explorations of consciousness.

Here is the main weakness in the philosophical methods standardly used in these cases: philosophers tend to assume that all the competences and dispositions that normal people exhibit regarding, say, colors, form a monolithic block, invulnerable to decomposition or dissociation into independent subcompetences and subdispositions. This handily excuses them from addressing the question of whether qualia are to be anchored to some subset or specific disposition. For instance, George Graham and Terry Horgan (2000) speak of "direct acquaintance with phenomenal character itself, acquaintance that provides the experiential basis for [a person's] recognitional/discriminatory capacities" (p. 73). If, to harken back to Wilfrid Sellars once again, qualia are what make life worth living, then qualia may *not* be the "experiential basis" for our ability to recognize colors from day to day, to discriminate colors, to name them. (I will explore this prospect from a different angle in the next chapter.)

The canonical first-person-accessibility or subjectivity of qualia is in trouble in any case, since, as change blindness demonstrates so vividly, one's first-person access to one's color qualia counts for nothing if it cannot be relied on to secure authority for simple judgments of the sort that elude people in these circumstances. Let me end this chapter by placing one more family of difficulties on the table for the believers in qualia: it has recently been demonstrated that many plants are sensitive to the ratio of red to infrared light reflected onto them, a measure of whether or not some green competitor is encroaching on their sunlight. When they sense that the neighborhood is getting green overhead, they adjust their own growth policy to invest more heavily to vertical growth, in order to compete more opportunistically. Now what is it like to be a plant surrounded by other green plants? Is it different from what it is like to be a green plant growing in splendid isolation? Is there anything it is like to be tree? Most of us, I suppose, will be inclined to answer in the negative, but if we then cast about for a reason for our judgment, there will be little to present. It just seems unlikely, I suppose, that plants have "feelings" or "subjectivity"—no matter how adroitly they utilize the spectral information falling on their surfaces. But then to preserve consistency we must forswear any support for our confidence that it is like something to be a bee, or a bat or a bird, that might derive from our appreciation of the intelligent use these animals make of their spectral information.

The concept of qualia is challenged on all sides by prospects that have simply not been considered by philosophers, and they need to forsake the cozy presumption of shared understanding with which they excuse themselves from the task of defining their term of art.

5   What RoboMary Knows

Frank Jackson's thought experiment about Mary the color scientist is a prime example of an intuition pump, a thought experiment that is not so much a formal argument as a little scenario, or vignette, that has been pumping philosophical intuitions with remarkable vigor since it first appeared in 1982. In fact, so much attention has it attracted over the years that *two* anthologies of Mariology are in preparation, celebrating and reviewing the twenty years that it has captivated philosophers' attention.[1]

> Mary is a brilliant scientist who is, for whatever reason, forced to investigate the world from a black-and-white room *via* a black-and-white television monitor. She specializes in the neurophysiology of vision and acquires, let us suppose, all the physical information there is to obtain about what goes on when we see ripe tomatoes, or the sky, and use terms like *red*, *blue*, and so on. She discovers, for example, just which wavelength combinations from the sky stimulate the retina, and exactly how this produces *via* the central nervous system the contraction of the vocal cords and expulsion of air from the lungs that results in the uttering of the sentence "The sky is blue" . . . What

1. A version of this chapter is forthcoming in one of those anthologies, Alter forthcoming. I am grateful to Diana Raffman, Bill Lycan, Victoria McGeer, and my students for many discussions, on email and in person, on the ins and outs of this argument.

will happen when Mary is released from her black-and-white room or is given a color television monitor? Will she *learn* anything or not? It seems just obvious that she will learn something about the world and our visual experience of it. But then it is inescapable that her previous knowledge was incomplete. But she had *all* the physical information. *Ergo* there is more to have than that, and Physicalism is false. ... (Jackson 1982, p. 128)

For sheer volume and reliability—twenty years without showing any signs of wear and tear—this must count as one of the most successful intuition pumps ever devised by analytical philosophers. But is it a good intuition pump? How could we tell? Douglas Hofstadter's classic advice (Hofstadter and Dennett 1981, p. 375) to philosophers confronted by a thought experiment is to treat it the way scientists treat a phenomenon of interest: vary it, turn it over, examine it from all angles, and in different settings and conditions, just to make sure you aren't taken in by illusions of causation. Turn all the knobs, he said, and see if the thing still pumps the same intuitions. This chapter, then, is an exercise in knob turning.

1 Mary and the Blue Banana

More than a decade ago, I conducted a preliminary exploration of the knobs and issued a verdict that has been almost universally disregarded: "Like a good thought experiment, its point is immediately evident even to the uninitiated. In fact it is a bad thought experiment, an intuition pump that actually encourages us to misunderstand its premises!" (1991, p . 398). In fact it is much more difficult to imagine the scenario correctly than people suppose, so they imagine something easier, and draw their conclusions from that mistaken base. In an attempt to bring out the flaws in the thought experiment, I encouraged people to consider a variant ending:

And so, one day, Mary's captors decided it was time for her to see colors. As a trick, they prepared a bright blue banana to present as her first color experience ever. Mary took one look at it and said "Hey! You tried to trick me! Bananas are yellow, but this one is blue!" Her captors were dumfounded. How did she do it? "Simple," she replied. "You have to remember that I know *everything*—absolutely everything—that could ever be known about the physical causes and effects of color vision. So of course before you brought the banana in, I had already written down, in exquisite detail, exactly what physical impression a yellow object or a blue object (or a green object, etc.) would make on my nervous system. So I already knew exactly what *thoughts* I would have (because, after all, the "mere disposition" to think about this or that is not one of your famous qualia, is it?). I was not in the slightest surprised by my experience of blue (what surprised me was that you would try such a second-rate trick on me). I realize it is *hard for you to imagine* that I could know so much about my reactive dispositions that the way blue affected me came as no surprise. Of course it's hard for you to imagine. It's hard for anyone to imagine the consequences of someone knowing absolutely everything physical about anything!" (1991, pp. 399–400)

It is standardly assumed without argument that things could not proceed this way. As Jackson disarmingly put it in the original article, "It seems just obvious that she will learn something about the world and our visual experience of it" (1982, p. 128). That, I claimed, is a mistake, and that is what is wrong with Mary as a thought experiment. It just feels so good to conclude that Mary has a revelation of *some* sort when she first sees color that nobody wants to bother showing that this is how the story *must* go. In fact it needn't go that way at all. My variant was intended to bring out the fact that, absent any persuasive *argument* that this could not be how Mary would respond, my telling of the tale had the same status as Jackson's: two little fantasies pulling in opposite directions, neither with any demonstrated authority. I thought that I had said enough to make my point, but a decade of further writing on Mary by many philosophers and their students has shown me that I should have been more patient, more explicit, in my objections. I underestimated the strange allure of this intuition pump by a wide margin. So I am returning to the fray, and this time I will make my case at a more deliberate pace, dotting the *is* and crossing the *ts*.

First, I have found that some readers—maybe most—just didn't get my blue banana alternative.[2] What was I saying? I was saying that Mary had figured out, using her vast knowledge of color science, *exactly what it would be like for her to see* something red, something yellow, something blue in advance of having those experiences.[3] I asserted this flat out—*in your face*, as it were—in order to expose to view the fact that people normally assume that this is impossible on the basis of no evidence or theory or argument, but just on the basis of ancient philosophical tradition going back at least to John Locke. Perhaps a little dialogue will help bring out the intended point:

TRAD: What on earth do you mean? *How* could Mary do that?

2. For instance, Howard Robinson (1993) supposes that I am illicitly helping myself to the premise that Mary knows "every particular physical thing that is going on" (p. 175), but my claim does not at all depend on such a strong claim, as will be clear from the variations I develop here.

3. Robinson (1993) also claims that I beg the question by not honoring a distinction he declares to exist between knowing *"what one would say and how one would react"* and knowing "what it is like." If there is such a distinction, it has not yet been articulated and defended, by Robinson or anybody else, so far as I know. If Mary knows *everything* about what she would say and how she would react, it is far from clear that she wouldn't know what it would be like.

DCD: It wasn't easy. She deduced it, actually, in a 4765-step proof (for red—once she'd deduced what red would look like to her, green fell into line with a 300-step lemma, and the other colors—and all the hues thereof—were relatively trivial extensions of those proofs).

TRAD: You're just making all that up! There are no such proofs!

DCD: This is a thought experiment; I get to make up all sorts of things. Can you prove that there are no such proofs? What established fact or principle am I contradicting when I help myself to a scenario in which she deduces what colors would look like to her from everything she knows about color?

TRAD: Look. It's just obvious! *You can't deduce what a color looks like if you've never seen one!*

DCD: That's an interesting folk theorem, I must say. Here's another: If you burp, sneeze, and fart all at the same time, you die. Sounds sort of plausible to me. But is there any scientific backing for either one of them?

2   "Surely" She'll Be Surprised

The Mary thought experiment might be intended simply to draw out and illustrate vividly the implications of a fairly standard way of thinking that many, probably most, people have. As such, it might be a useful anthropological exercise, an investigation of folk psychology laid bare, as I noted in chapter 2. But those who have championed Mary have thought that it might actually prove something bigger, not just the conclusion that most people's unexamined assumptions imply dualism—I think we already knew that, but maybe not—but the conclusion that dualism is true! The fact that philosophers would so much as *entertain* such an interpretation of such a casual exercise of the imagination fills me with astonishment. I had no idea philosophers still put so much faith in the authority of their homegrown intuitions. It is almost as if one thought one could prove that the Copernican theory was false by noting that it "seems just obvious" that the Earth doesn't move and the Sun does.

In a recent article, "Mary Mary Quite Contrary," George Graham and Terry Horgan (henceforth G&H) have usefully managed to distill precisely the unargued intuition that I have been attempting to isolate and discredit for fifteen years or more—the one we might express as "She'll be surprised, dammit!" G&H begin by distinguishing two main materialist responses to Mary: thin and thick materialism. Thin materialism, of which I am one of the few exponents, denies that Mary learns anything after release. Thick materialists attempt to salvage materialism while going along with the gag that Mary is startled, delighted, surprised, or something like that, when she is released from her colorless captivity. G&H's strategy is first to declare briskly that thin materialism is a nonstarter in need of no refutation since it "has been amply criticized by others" (p. 63). The only critics they list are McConnell (1994) and Lycan (1996). Since I replied at some length to McConnell in the same journal (Dennett 1994a), and since Lycan doesn't criticize my version of thin materialism, I don't find this criticism ample, but I must admit that G&H are only going along with the mainstream in ignoring my brand of thin materialism. That's why the current essay is necessary.

G&H spell out the best of the thick materialist campaigns—Michael Tye's PANIC—and imagine their own variation on the original theme: Mary Mary, the daughter of the original Mary, and a devotee of Tye's brand of thick materialism. According to Tye's PANIC theory, "phenomenal character is one and the same as Poised Abstract Nonconceptual Intentional Content" (Tye 1995, p. 137), which means roughly that it is content that is "in position to make a direct impact on the belief/desire system" and is about non-concrete, non-conceptualized discriminable properties. It follows from Tye's view, they claim, that Mary Mary *shouldn't be surprised*. As they say, "In the end, Tye's version of thick materialism is just *too thin*. And this problem threatens to arise for any materialist treatment of phenomenal content" (p. 77).

I had previously viewed Tye's alternative to my brand of thin materialism as giving too much ground to the qualophiles, the lovers of phenomenal content, but thanks now to G&H I can welcome him into my underpopulated fold as a thin materialist *malgré lui*, someone who has articulated much more painstakingly than I had just what sorts of functionalistically explicable complexities go to *constitute* the what-it-is-likeness, the so-called *phenomenality*, of conscious experience. I applaud G&H's analysis of Mary Mary's predicament, leading inexorably to the conclusion that since she already knows all the facts, has all the information needed to have anticipated *all* the noticeable, remarkable-upon properties of her debut experience in a colored world, she should not, in spite of what Tye claims, be (or expect to be) surprised. Here, in a nutshell, is what they say:

> First, what is psychologically significant about the PANIC properties is just the functional/representational role they play in human cognitive economy—something that Mary thoroughly understands already, by virtue of her scientific omniscience. . . . Second, what is psychologically significant about phenomenal concepts (given Tye's theory) is that they are *capacity-based* concepts. . . . But she already

understands these capacities thoroughly, including how PANIC states subserve them, even though she does not possess the capacities herself. No expected surprises there, either.

Third, the psychological distinctiveness of beliefs and knowledge-states employing phenomenal concepts is completely parasitic (given Tye's theory) upon the capacity-based nature of the phenomenal concepts. So she already understands well the *nature* of these beliefs and knowledge-states. . . . So Mary Mary, as a True Believer in Tye's PANIC theory of phenomenal consciousness, has no good reason to expect surprise or unanticipated delight upon being released from her monochrome situation. (G&M, pp. 71–72)

In short, Tye should join me in predicting that Mary Mary, like her mother Mary, would *not* be surprised or delighted at all. She's been there, done that, in her vast imagination already, and has nothing left to learn. So what's the problem? Why don't G&H join Tye and me? (I'm presuming for the fun of it that Tye is now on my side.) Because—and here comes the superpure, double-distilled intuition that I've been gunning for—"Surely, we submit, she should be both surprised and delighted" (p. 72). "Surely." As I noted in "Get Real" (Dennett 1994a) in one of my many commentaries on Ned Block, "Wherever Block says "Surely," look for what we might call a mental block" (p. 549). Block is perhaps one of the most profligate abusers of the "surely" operator among philosophers, but others routinely rely on it, and every time they do a little alarm bell should ring. Here is where the unintended sleight-of-hand happens, whisking the false premise by the censors with a nudge and a wink. G&H do pause momentarily to ask why they are so sure, and this is what they answer (p. 72):

What will surprise and delight Mary Mary . . . is (it seems to us) the unanticipated *experiential basis* of her concept-wielding, recognitional/discriminatory, capacities and the acknowledged richness of her experience; she never expected polychromatic experience to be like *this*.

I know that it seems to many people that there is this extra "richness," this *"experiential basis"* over and above all the PANIC details, but I have claimed that they are just wrong about this, and I have offered a diagnosis of the sources of this deep-seated theorists' illusion. In "Quining Qualia" (Dennett 1988), I discussed the example of the torn Jell-O box, half of which has shape property M, and the other half of which is the only *practical* M-detector: the shape may *defy description* but it is not literally ineffable or unanalyzable; it is just extremely rich in information. It is a mistake to inflate practical indescribability into something metaphysically more portentous, and I have been urging people to abandon this brute hunch, tempting though it may be. G&H cannot bring themselves to abandon the intuition, but more important, they cannot even bring themselves to acknowledge that their whole case thus comes down to simply announcing their continued allegiance to a claim that, whether it is true or false, has been declared false and hence could use some support. They offer no support for it, but they do keep coming back to it, again and again:

> Although phenomenal states may indeed play a PANIC role in human psychological economy, their phenomenal character is not reducible to that role. It is something more, something surprising and delightful. (G&H, p. 73)

Who says? This is just what I have denied, at length.

> Its greater richness is what is surprising and delightful about it, and Tye's theory leaves this out. (Ibid.)

This "greater richness" is just what needs to be demonstrated, not assumed. After all, the point of the Mary example is supposed to be that although thanks to her perfect knowledge she can anticipate *much* of what it will be like to see colors, she cannot anticipate it *all*. Since some of us have claimed that there is no reason to deny that all the "greater richness" *is* accessible to Mary in advance, this bald assertion by G&H is question-begging. It simply won't do to lean on the obvious fact that under normal circumstances, indeed under any circumstances except the wildly improbable extreme circumstances of this thought experiment, Mary would learn something.

> But she *will* experience surprise and unanticipated delight, upon release from her monochromatic environment—which presumably should lead her to repudiate the materialist theory she previously accepted. (G&H, p. 74)

So they say. Now thin materialism may, in the end, be false, but you can't argue against it by just saying "Surely not!" I have claimed that the richness we appreciate, the richness that we rely on to anchor our acts of inner ostension and recognition is *composed of* and *explained by* the complex set of dispositional properties that Tye has called PANIC properties. G&H make the mistake of assuming that there is, in addition to all this, a layer of "direct acquaintance" with "phenomenal properties." They say baldly:

> There is also direct acquaintance with phenomenal character itself, acquaintance that provides the experiential basis for those recognitional/discriminatory capacities. (G&H, p. 73)

And also:

> She claims to be delighted. . . . Auto-phenomenology suggests strongly, *very* strongly, that she is right about this: the intrinsic phenomenal character of color experience is distinct from, and provides the basis for, these recognitional/discriminatory capacities. (G&H, p. 77)

As we saw in chapter 4, this is just about backwards. These capacities are themselves the basis for the (illusory) belief that one's experience has "intrinsic phenomenal character," and we first-persons have no privileged access at all into the workings of these capacities. That, by the way, is why we shouldn't do autophenomenology. It leads us into temptation: the temptation to take our own first-person convictions not as data but as the undeniable truth.

> So on his [Tye's] story, Mary Mary's post-release heterophenomenological claims evidently must be viewed as rationally inappropriate, and thus as embodying some kind of error or illusion. *That* is the basic problem: the apparent failure to provide adequate theoretical accommodation for the manifest phenomenological facts. (G&H, p. 77)

The basic problem, they say, is dealing with these "manifest" facts, but it's only a problem if, in fact, she will learn something. It is not a problem for my view (and Tye's, if he'll join the thin materialists); she won't learn anything, and she won't be surprised; there are no such manifest phenomenological facts. At this point, if you are like many of my students, you are beset with frank incredulity. *Of course* Mary learns something on release! She *has to*! Oh? Then please give me an argument, based on premises we can all accept, that demonstrates this. But I have never seen such an argument even attempted. "It stands to reason!" people say, and then they decline to offer any reasons, thinking them somehow uncalled for. I call for them.

In response to the previous paragraph in an earlier draft, Bill Lycan has answered the call:

> Here's a way to see why some of us think Mary does learn something. What one knows when one knows w.i.l. [what it's like] to experience a blue sensation is ineffable; at least, it's very tough to put into (noncomparative) words. One resorts to the frustrated demonstrative: "It's

like . . . *this.*" The reason physically omniscient Mary doesn't know what it's like is that the ineffable and/or the ineliminably demonstrative can't be deduced, or even induced or abduced, from a body of impersonal scientific information. (Personal communication)

I daresay that Lycan speaks for many who are sure that Mary learns something, so now we have an explicit rendering of a background presumption of ineffability and an illustration of the role it plays in the argument I call for. Now what about that argument? First of all, nobody could deny that these propositions ventured by Lycan are large theoretical claims, not minimal logical intuitions or the immediate, unvarnished judgments of experience. *What one knows when one knows what it's like to experience a blue sensation is ineffable.* I suppose the concept of ineffability being appealed to here would get elaborated along these lines:

It is not the case that there is a string of demonstrative-free sentences of natural language, of any length, that adequately expresses the knowledge of what it is like to experience a blue sensation.

One would like to see that proved. (I'm being ironic. Of all the things one might want to construct a formal theory of, *ineffability* is way down the list, but it might be worthwhile to consider the difficulty of any such undertaking.) Presumably one wants to contrast the ineffability of what it's like to experience a blue sensation with, say, the ready effability (if I may) of what it's like to experience a triangle. Someone who has never seen or touched a triangle can presumably be told in a few well-chosen words just what to expect, and when they experience their first triangle, they should have no difficulty singling it out as such on the basis of the brief description they had been given.

They will learn nothing. With blue and red it is otherwise—that, at any rate, is the folk wisdom relied on by Jackson's example. (He wouldn't have gotten far with a thought experiment about Mary the geometer who was prevented from seeing or touching triangles.) But if what it is like to see triangles can be adequately conveyed in a few dozen words, and what it is like to see Paris by moonlight in May can be adequately conveyed in a few thousand words (an empirical estimate based on the variable success of actual attempts by novelists), are we really so sure that what it is like to see red or blue can't be conveyed to one who has never seen colors in a few million or billion words? What is it about the experience of red, or blue, that makes this task impossible? (And don't just say: they're *ineffable*.)

We are enjoined by the extremity of the thought experiment to take this question seriously. Remember, Mary knows *everything* about color that can be learned by physical science, and she presumably has the attention span and powers of comprehension required to handle ten-billion-word treatises on what it is like to see red as easily as twenty-five-words-or-less on triangles. Lycan says "at least it's very tough to put into (noncomparative) words," but this is not a thought experiment about difficulty; it's a thought experiment about impossibility. The fact that people find it hard to imagine that any description of what it's like to see red could do the job is negligible support. Faced with such a formidable task, one does indeed fall back on what Lycan aptly calls the "frustrated demonstrative," but it is a long way from the undeniable claim that it is *very tough* to think of ways of characterizing what it is like without resorting to such private demonstratives, to the grand claim that such private demonstratives are strictly speaking ineliminable. And only absolute ineliminability would carry any weight in an argument against the *possibility* of Mary inferring what it would be like for her to see red.

So I stick to my guns. The standard presumption that Mary learns something, that Mary *could not* have figured out just what it would be like for her to see colors, is a bit of folk psychology with nothing but tradition—so far—in its favor. (This is an invitation to philosophers to call my bluff and construct an argument that shows, from unproblematic shared premises, that Mary cannot figure out what specific colors will look like to her.)

3 You Had to Be There!

Another unargued intuition exploited covertly by the Mary intuition pump comes in different varieties, all descended inauspiciously from Locke and Hume (think of Hume's missing shade of blue). This is the idea that the "phenomenality" or "intrinsic phenomenal character" or "greater richness"—whatever it is—cannot be constructed or derived out of lesser ingredients. Only actual experience (of color, for instance) can lead to the knowledge of what that experience is like. Put so boldly, its question-beggingness stands out like a sore thumb, or so I once thought, but apparently not, since versions of it still get articulated. Here are two, drawn from Tye and Lycan:

> Now, in the case of knowing via phenomenal concepts, knowing what it is like to undergo a phenomenal state type *P* demands the capacity to represent the phenomenal content of *P* under those concepts. But one cannot possess a predicative phenomenal concept unless one has actually undergone token states to which it applies. (Tye 1995, p. 169)[4]

4. Earlier Tye had noted simply that "possessing the phenomenal concept *red* requires that one have experienced red . . . possession consists (very roughly) in having available a state that has a causal history that links it with the relevant experiences . . ." (p. 167).

As Nagel emphasizes, to know w.i.l., one must either have had the experience oneself, in the first person, from the inside, or been told w.i.l. by someone who has had it and is psychologically very similar to oneself. (Lycan forthcoming)[5]

The role of this presupposition is revealed in the many attempts in the literature to guarantee that Mary doesn't cheat, somehow smuggling the experience of color into her cell. What special care must be taken to prevent Mary from taking surreptitious sips from the well of color? Jackson supposes that Mary is obliged to wear white gloves at all times, and no mirrors are in her cell, and so on, but these blockades erected by Jackson in his original telling have long been recognized as insufficient as they stand. For instance, Mary might innocently rub her closed eyes one day and create some colored "phosphenes" (try it—I just got a nice deep indigo one right in the middle of my visual field). Or she might use her vast knowledge to engage in some transcranial magnetic stimulation of her color-sensitive cortical regions, producing even gaudier effects for her to sort out. Should a sophisticated alarm system be installed in her brain, to cut short any dream "in color" that she might innocently wander into by happenstance? Is it in fact possible for a person to dream in color if that person has never seen colors while awake? (What do you think? Some might be tempted to respond: "Naw. The colors have to *get in there* through open eyes in order to be available for later use in dreaming." That's the

5. Lycan's second alternative is a startling concession, unremarked upon. If one can be told, why can Mary not simply be told, by standard color-enjoyers psychologically very similar to her? I don't think Nagel allows this, and if Lycan does, the game is over. Thin materialism wins. In the context, Lycan is imagining the variation in which Mary is born color-blind, which explains the lapse: those who could tell her are not "psychologically very similar" to her.

Lockean premise laid bare, and presumably nobody would be seduced by it in such a raw form today.) The thought experiment tacitly presupposes that in spite of her impoverished visual environment, Mary's color vision system is still intact. In fact, this is a nontrivial empirical assumption, given what is known about the ready reassignment of unused cortical resources in other regards, but it is also unrealistic in its assumption that you can eliminate all spectral information in the light by just painting everything white (which shade of white?) and black (Akins 2001). Setting these actually important empirical complications aside for the sake of argument, we are supposed to agree that she already has "in there" everything she needs to experience color; it just hasn't been stimulated. A dream could trigger the requisite activity as readily, presumably, as any external stimulus to the open eyes. There are no doubt myriad ways of short-circuiting the standard causal pattern and producing color experience in the absence of external-world color.

More ominously for the prospects of the thought experiment, there are no doubt myriad ways of adjusting the standard causal pattern to produce some state of the brain that is *almost the same as* the sort of state that underlies standard color experience, but that differs in ways that are crucial to the clarity of the scenario, and to what it is meant to prove. What started out as a crisp, clean, "intuitive" predicament is being pulled out of shape by the inconvenient complications of science. According to the original thought experiment, it is the subjective, internal *experience of color*, however produced, that is held to be a prerequisite for knowing what it is like to see red, but now that we recognize that there are paths to such experience other than that of the standard eyes-open-and-awake, the experience of color cannot so readily be distinguished from other states of mind that have many of the effects of experiences of color without clearly being experiences of color. What, for instance, is the difference between imagining you are experiencing red and experiencing red? If you actually succeed in imagining you are experiencing red, do you *thereby* succeed in experiencing red? If so, the circularity of the thought experiment looms. We are told that Mary in her cell can't imagine what it's like to experience red, try as she might.

But suppose she doesn't accept this limitation and does try her best, cogitating for hours on end, and one day she tells us she just got lucky and succeeded. "Hey," she says, "I was just daydreaming, and I stumbled across what it's like to see red, and, of course, once I noticed what I was doing I tested my imagination against everything I knew, and I confirmed that I had, indeed, imagined what it's like to see red!" Doubting her, we test her by showing her a display of three differently colored circles, and she immediately identifies the red one as red. What would we conclude?

A. Jackson was wrong; Mary *can* figure out what it's like to see red in the absence of any experience of red; or

B. Mary didn't *figure out* what it is like to see red; she had to resort to (highly intelligent, theory-guided) exercises of *imagining* in order to come to know what it is like to see red. By *imagining* red, she was actually illustrating Jackson's point. As her example shows, you can't know what it's like before you've actually experienced what it's like.

This is an awkward moment: a simple variation on the tale that clearly refutes it or clearly vindicates it, depending on how you interpret what happened. If B is the only conclusion Jackson intended, then we philosophers have been spending a lot of time and energy on what appears in retrospect to be a relatively trivial definitional issue: nothing is going to be allowed to *count* as knowing what it's like to see red without also counting as an experience of red. This embarrassing outcome just couldn't arise, many philosophers think, because they are so certain that Mary just could not accomplish this feat. I insist otherwise.

Before looking more closely at this contretemps, let's consider one other variation, one I would have thought was the obvious variation for philosophers: Swamp Mary.[6] Suppressing my gag reflex and my giggle reflex, here she is:

> *Swamp Mary:* Just as standard Mary is about to be released from prison, still virginal with regard to colors and aching to experience "the additional and extreme surprise, the unanticipated delight, or the utter amazement that lie in store for her" (G&H, p. 82), a bolt of lightning rearranges her brain, putting it by Cosmic Coincidence into exactly the brain state she was just about to go into *after* first seeing a red rose. (She is left otherwise unharmed of course; this is a thought experiment.) So when, a few seconds later, she is released, and sees for the first time, a colored thing (that red rose), she says just what she would say on seeing her *second* or *nth* red rose. "Oh yeah, right, a red rose. Been there, done that."

Let me try to ensure that the point of this variation is not lost. I am *not* discussing the case in which the bolt of lightning

6. Gabriel Love suggested this hybrid to me, and I think he was inspired by considering another: In "The Trouble with Mary," Victoria McGeer (2003) has written persuasively and elegantly on what happens when you hybridize Mary and zombies. I have expressed my distrust of all such thought experiments in Dennett 1994a; excerpted (with revisions) in Dennett 1996a.

gives Swamp Mary a hallucinatory experience of a red rose. That is, of course, one more "possibility," but it is not the possibility I am introducing. I am supposing instead that the bolt of lightning puts Swamp Mary's brain into the dispositional state, the competence state, that an experience of a red rose *would have put her brain into* had such an experience (hallucinatory or not) occurred. So, after her Cosmic Accident, Swamp Mary may *think* that she's seen a red rose, experienced red, been in a token brain state of the type that subserves experiences of red, but she hasn't. It's just as if she had. Maybe she *wrongly remembers or seems to remember* (just like Swampman—see Davidson 1987) having seen a red rose, or maybe, in spite of her lacking any such episodic memories, her competences are otherwise all *as if* she had had such episodes in her past. (After all, you could forget your first color experiences and still have phenomenal concepts, couldn't you?) Ex hypothesi she didn't have any such experiences, whatever she now thinks; any bogus memories of color were inserted illicitly in her memory box by the lightning bolt. Hey, (surely) it's *logically* possible. Swamp Mary is exactly like Mary, an atom-for-atom duplicate of Mary at every moment of her life except for a brief interlude of lightning that performs the accidental (but not supernatural) feat of doing in a flash exactly what Mary's looking at the rose would do by more normal causal routes. It follows that those who think "that there are certain concepts that . . . can only be possessed and deployed on the basis of having undergone the relevant conscious experiences oneself" (G&H, speaking of Tye, p. 65) may be right as a matter of contingent fact, but it is logically possible for one to acquire this enviable ability by accidental means. (These words stick in my throat, but I'm playing the game as best I can.)

We now have two routes to Mary's post-release knowingness: the Approved Path of "undergoing the relevant conscious experiences oneself" and the logically possible Cosmic Accident Path taken by Swamp Mary. The second path is a throwaway, not worth discussing. What *is* worth discussing is a third route to this summit, not a pseudo-miracle but an ascent by good hard work: Mary puts all her scientific knowledge of color to use and *figures out* exactly what it is like to see red (and green, and blue) and hence is not the least bit surprised when she sees her first rose. Since some philosophers think that Mary just could not do this, let me attempt to show just how she could, helping myself to a little simplification: in my final twist of the knob, I am going to turn Mary the color scientist into a robot.

4 RoboMary

I will begin with a deliberately simpleminded version, for clarity, and gradually add the complications that the disbelievers insist on. In the spirit of cooperative reverse-engineering, I'm numbering the knobs on my intuition pump, and adding comments on how the knob settings agree or differ from other models of the basic intuition pump.

1. RoboMary is a standard Mark 19 robot, except that she was brought online without color vision; her video cameras are black-and-white, but everything else in her hardware is equipped for color vision, which is standard in the Mark 19.

So, just like Mary, RoboMary's *internal* equipment is "normal" for color "vision" but she is being peripherally prevented from getting the appropriate input temporarily—from "birth." RoboMary's black-and-white cameras stand in nicely for the isolation of human Mary, and we can let her wander at will through the psychophysics and neuroscience journals reading with her black-and-white-camera eyes.

> 2. While waiting for a pair of color cameras to replace her black-and-white cameras, RoboMary learns everything she can about the color vision of Mark 19s. She can even bring colored objects into her prison cell along with normally color-sighted Mark 19s and compare their responses—internal and external—to hers.

This was something that Mary could do, of course, only somewhat more tediously—she had to watch black-and-white TV while conducting all the experiments she'd need to get that admirably complete compendium of physical information. This suggests a modest improvement that could be made in Jackson's original experiment, in which Mary's eyes—and especially, the cones in her retinas—are declared normal, and the entire color-blockade has to be accomplished with prison walls, confiscation of mirrors, white gloves, and so on. As various commentators have observed, such a world would still be an ample source of chromatic input—shadows and the like, not to mention the different shades of "white." It would have been a lot cleaner for Jackson's original telling if he had just stipulated that Mary had had a pair of camcorders with black-and-white eyepieces strapped over her eyes, peering at the world all her life like somebody videotaping her vacation in Europe.

> 3. She learns all about the million-shade color-coding system that is shared by all Mark 19s.

We don't know that human beings share the same color-coding system. In fact we can be quite certain they don't, but so what? This is just a complication; if Mary knows *everything*, she knows all the variations of human color-coding, including her own.

4. Using her vast knowledge, she writes some code that enables her to colorize the input from her black-and-white cameras (à la Ted Turner's cable network) according to voluminous data she gathers about what colors things in the world are, and how Mark 19s normally encode these. So now when she looks with her black-and-white cameras at a ripe banana, she "sees it as yellow" since her colorizing prosthesis has swiftly looked up the standard ripe-banana color-number-profile and digitally inserted it in each frame in all the right pixels.

So now she sees a ripe banana as yellow? Isn't this simply the robot version of phosphenes and transcranial magnetic-stimulation, cheating ways of getting color experience into RoboMary? Or is it simply a way of dramatizing the immense knowledge of color "physiology" that RoboMary, like Mary, enjoys? What is either of them allowed to do with their knowledge? Let's turn the knob both ways, and see what happens. In this first, and simplest, setting, we declare that just as Mary is entitled to *use her imagination* in any way she likes in her efforts to come up with an anticipation of what it's going to be like to see colors, RoboMary is entitled to use her imagination, and that is just what she is doing—after all, no hardware additions are involved: she is just considering, by stipulation, what it might be like under various conditions to see colors (we can suppose she considers dozens of variant colorization codings).

5. She wonders if the ersatz coloring scheme she's installed in herself is high fidelity. So during her research and development phase, she checks the numbers in her registers (the registers that transiently store the information about the colors of the things in front of her cameras) with the numbers in the same registers of other Mark 19s looking at the same

objects with their color-camera-eyes, and makes adjustments when necessary, gradually building up a good version of normal Mark 19 color vision.

In the case of RoboMary it is obvious what sorts of use she can make of her knowledge about color and color vision in Mark 19s. It is far from obvious, of course, how Mary could make use of her knowledge. But that just shows how treacherous the original intuition pump is; it discourages us from even trying to imagine the task facing Mary, the task of figuring out what it is like to see red.

> 6. The big day arrives. When she finally gets her color cameras installed, and disables her colorizing software, and opens her eyes, she notices . . . nothing. In fact, she has to check to make sure she has the color cameras installed. She has learned nothing. She already knew exactly what it would be like for her to see colors.

Before turning to the variation that prohibits RoboMary from adjusting her color registers in this way (thereby producing in herself premature color experiences), I must consider what many will view as a more pressing objection:

> Robots don't *have* color experiences! Robots don't have *qualia*! This scenario isn't remotely on the same topic as the story of Mary the color scientist.

I suspect that many will want to endorse this objection, but they really must restrain themselves, on pain of begging the question most blatantly. Contemporary materialism—at least in my version of it—cheerfully endorses the assertion that *we* are robots of a sort—made of robots made of robots. Those who rule out my scenario as irrelevant from the outset are not arguing for the falsity of materialism; they are assuming it, and they illustrate that assumption in their version of the Mary story (interesting as anthropology, perhaps, but unlikely to shed any light on the science of consciousness).

5 Locked RoboMary

Now let's turn the knob and consider the way RoboMary must proceed if she is prohibited from tampering with her color-experience registers. I don't know how Mary could be crisply rendered incapable of using her knowledge to put her own brain into the relevant imaginative and experiential states, but I can easily describe the software that will prevent RoboMary from doing it. In order to prevent this sort of self-stimulation skullduggery, we arrange for RoboMary's color-vision system—the array of registers that transiently hold the codes for each pixel in Mary's visual field, whether seen or imagined—to be restricted to gray-scale values. This is simple: We arrange to code the gray-scale values (white through many grays to black) with numbers below a thousand, let's say, and simply filter out (by subtraction) any values for chromatic shades in the million-shade subjective spectrum of Mark 19s—and we put unbreakable security on this subroutine. Try as she might, RoboMary can't jigger her "brain" into any of the states of normal Mark 19 color vision. She has all her hard-won knowledge of that system of color vision, but she can't use it to adjust her own registers so that they match those of her conspecifics.

This doesn't faze her for a minute, however. Using a few terabytes of spare (undedicated) RAM, she builds a model of herself and *from the outside, just as she would if she were building a model of some other being's color vision,* she figures out just how she would react in every possible color situation.

I find people have trouble imagining just how intimate and vast this "third-person" knowledge would be, so we might indulge in a few details, to illustrate. She obtains a ripe tomato and plunks it down in front of her black-and-white cameras, obtaining some middling gray-scale values, which lead her into a variety of sequel states. She automatically does the usual "shape from shading" algorithm, obtaining normal convictions about the bulginess and so forth, and visually guided palpation gives her lots of convictions about its softness. She consults an encyclopedia about the normal color range of tomatoes, and she knows that these gray-scales in these lighting conditions are consistent with redness, but of course nothing comes to her directly about color, since she has black-and-white cameras, and moreover, she can't use her book-learning to adjust these values, since her color system is locked. So, as advertised, she can't put herself directly into the *red-tomato-experiencing* state, or even into the *red-tomato-imagining* state. She looks at the (gray-appearing) tomato and reacts however she does, in, say, hundreds or thousands of temporary settings of her cognitive machinery. (Researchers seldom direct their attention to more than one or two of the sequelae of a perceptual state they have managed to induce in their subjects. They tend to ignore what I have called the Hard Question—And Then What Happens? [Dennett 1991, p. 255]. A lot happens.) Call the voluminous state of her *total response* to the locked color-state *state A*. Then she compares state A with the state that her model of herself goes into. Her model isn't locked; it readily goes into the state that any normal Mark 19 would go into when seeing a red tomato. And, since this is her model of herself, it then goes into state B, the state she would have gone into if her color system hadn't been locked. RoboMary notes all the differences between state A, the state she was thrown into by her locked color system, and state B, the state she would have been thrown into had her color system not been locked, and—being such a clever, indefatigable and nearly omniscient being—makes all the necessary adjustments and *puts herself into state B*. State B is, by definition, *not* an illicit state of color-experience (or even color-imagination); it is the state that such an illicit state of color-experience normally causes (in a being just exactly like her). But now she can know just what it is like to see a red tomato, because she has managed to put herself into just such a dispositional state—this is of course the hard-work analogue of the miraculous feat wrought by the Cosmic Accident of the lightning bolt in the case of Swamp Mary.

Her epistemic situation when she has completed this vast labor is indistinguishable from her epistemic situation in which we allow her to colorize her actual input. There are no surprises for her when her color system is unlocked and she's given color cameras. In fact, when she completes her model of herself, down to the very last detail, she can arrange for it to take over for her locked onboard color system, a spare color system she can use much as the fictional Dennett uses his spare computer brain in "Where Am I?" (Dennett 1978, ch. 17). Remember: RoboMary knows all the physical facts, and that's a lot.

Finally, I find that some philosophers think that my whole approach to qualia is not playing fair: I don't respect the standard rules of philosophical thought experiments. "But Dan, your view is so *counterintuitive*!" No kidding. That's the whole point. Of course it is counterintuitive; nobody ever said that the true materialist theory of consciousness should be blandly intuitive. I have all along insisted that it may be *very* counterintuitive. That's the trouble with "pure" philosophical method here. It has no resources for developing, or even taking seriously, counterintuitive theories, but since it is a very good bet that the true materialist theory of consciousness will be highly counterintuitive (like the Copernican theory—at least at first), this means that "pure" philosophy must just blind itself to the truth and retreat into conservative conceptual anthropology until the advance of science puts it out of its misery. Philosophers have a choice: they can play games with folk concepts (ordinary language philosophy lives on, as a kind of aprioristic anthropology) or they can take seriously the claim that some of these folk concepts are illusion-generators. The way to take that prospect seriously is to *consider* theories that propose revisions to those concepts.

1 Clawing Our Way toward Consensus

As the Decade of the Brain (declared by President Bush in 1990) comes to a close, we are beginning to discern how the human brain achieves consciousness.[1] Dehaene and Naccache (2001— all 2001 citations below are to papers in this volume) see convergence coming from quite different quarters on a version of the global neuronal workspace model. There are still many differences of emphasis to negotiate, and, no doubt, some errors of detail to correct, but there is enough common ground to build on. I agree, and will attempt to rearticulate this emerging view in slightly different terms, emphasizing a few key points that are often resisted, in hopes of precipitating further consolidation. (On the eve of the Decade of the Brain, Baars [1988] had already described a "gathering consensus" in much the

1. This chapter originally appeared as the closing overview of the essays in a special issue of *Cognition* (2001) on the cognitive neuroscience of consciousness, edited by Stanislas Dehaene (reprinted in 2002 by The MIT Press). Several short passages in this chapter appear verbatim or nearly so in earlier chapters, but I have left them intact to preserve the context in which they originally were published.

same terms: consciousness, he said, is accomplished by a "distributed society of specialists that is equipped with a working memory, called a *global workspace*, whose contents can be broadcast to the system as a whole" [p. 42]. If, as Jack and Shallice [2001] point out, Baars's functional neuroanatomy has been superceded, this shows some of the progress we've made in the intervening years.)

A consensus may be emerging, but the seductiveness of the paths not taken is still potent, and part of my task here will be to diagnose some instances of backsliding and suggest therapeutic countermeasures. Of course those who still vehemently oppose this consensus will think it is I who need therapy. These are difficult questions. Here is Dehaene and Naccache's (2001) short summary of the global neuronal workspace model, to which I have attached some amplificatory notes on key terms, intended as friendly amendments to be elaborated in the rest of the paper:

> At any given time, many modular (1) cerebral networks are active in parallel and process information in an unconscious manner. An information (2) becomes conscious, however, if the neural population that represents it is mobilized by top-down (3) attentional amplification into a brain-scale state of coherent activity that involves many neurons distributed throughout the brain. The long distance connectivity of these "workplace neurons" can, when they are active for a minimal duration (4), make the information available to a variety of processes including perceptual categorization, long-term memorization, evaluation, and intentional action. We postulate that this global availability of information through the workplace is (5) what we subjectively experience as a conscious state.

> (1) Modularity comes in degrees and kinds; what is being stressed here is only that these are specialist networks with limited powers of information processing.

(2) There is no standard term for an event in the brain that carries information or content on some topic (e.g., information about color at a retinal location, information about a phoneme heard, information about the familiarity or novelty of other information currently being carried, etc.). Whenever some specialist network or smaller structure makes a discrimination, fixes some element of content, "an information" in their sense comes into existence. "Signal," "content-fixation" (Dennett 1991), "micro-taking" (Dennett and Kinsbourne 1992), "wordless narrative" (Damasio 1999), and "representation" (Jack and Shallice 2001) are among the near-synonyms in use.

(3) We should be careful not to take the term "top-down" too literally. Since there is no single organizational summit to the brain, it means only that such attentional amplification is not just modulated "bottom-up" by features internal to the processing stream in which it rides, but also by *sideways* influences, from competitive, cooperative, collateral activities whose emergent net result is what we may lump together and call top-down influence. In an arena of opponent processes (as in a democracy) the "top" is distributed, not localized. Nevertheless, among the various competitive processes, there are important bifurcations or thresholds that can lead to strikingly different sequels, and it is these differences that best account for our pretheoretical intuitions about the difference between conscious and unconscious events in the mind. If we are careful, we can use "top-down" as an innocent allusion, exploiting a vivid fossil trace of a discarded Cartesian theory to mark the real differences that that theory misdescribed. (This will be elaborated in my discussion of Jack and Shallice 2001 below.)

(4) How long must this minimal duration be? Long enough to make the information available to a variety of processes—

that's all. One should resist the temptation to imagine some *other* effect that needs to build up over time, because . . .

(5) The proposed consensual thesis is not that this global availability *causes* some further effect or a different sort altogether—igniting the glow of conscious qualia, gaining entrance to the Cartesian Theater, or something like that—but that it *is*, all by itself, a conscious state. This is the hardest part of the thesis to understand and embrace. In fact, some who favor the rest of the consensus balk at this point and want to suppose that global availability must somehow kindle some special effect over and above the merely computational or functional competences such global availability ensures. Those who harbor this hunch are surrendering just when victory is at hand, I will argue, for these "merely functional" competences are the very competences that consciousness was supposed to enable.

Here is where scientists have been tempted—or blackmailed—into defending unmistakably *philosophical* theses about consciousness, on both sides of the issue. Some have taken up the philosophical issues with relish, and others with reluctance and foreboding, with uneven results for both types. In this paper I will highlight a few of the points made and attempted, supporting some and criticizing others, but mainly trying to show how relatively minor decisions about word choice and emphasis can conspire to mislead the theoretician's imagination. Is there a "Hard Problem" (Chalmers 1995, 1996), and if so what is it, and what could possibly count as progress toward solving it? Although I have staunchly defended—and will defend here again—the verdict that Chalmers's "Hard Problem" is a theorist's illusion (Dennett 1996c, 1998a)—something inviting therapy, not a real problem to be solved with revolutionary new science—I view my task here to be dispelling confusion first, and taking sides second. Let us see, as clearly as we can, what the question is, and is not, before we declare any allegiances.

Dehaene and Naccache (2001) provide a good survey of the recent evidence in favor of this consensus, much of it analyzed in greater deal in the other papers in that volume, and I would first like to supplement their survey with a few anticipations drawn from farther afield. The central ideas are not new, though they have often been overlooked or underestimated. In 1959, the mathematician (and coiner of the term "artificial intelligence") John McCarthy, commenting on Oliver Selfridge's pioneering Pandemonium, the first model of a competitive, nonhierarchical computational architecture, clearly articulated the fundamental idea of the global workspace hypothesis:

> I would like to speak briefly about some of the advantages of the pandemonium model as an actual model of conscious behaviour. In observing a brain, one should make a distinction between that aspect of the behaviour which is available consciously, and those behaviours, no doubt equally important, but which proceed unconsciously. If one conceives of the brain as a pandemonium—a collection of demons—perhaps what is going on within the demons can be regarded as the unconscious part of thought, and what the demons are publicly shouting for each other to hear, as the conscious part of thought. (McCarthy 1959, p. 147)

And in a classic paper, psychologist Paul Rozin argued that

> specializations . . . form the building blocks for higher level intelligence. . . . At the time of their origin, these specializations are tightly wired into the functional system they were designed to serve and are thus inaccessible to other programs or systems of the brain. I suggest that in the course of evolution these programs become more *accessible* to other systems and, in the extreme, may rise to the level of consciousness and be applied over the full realm of behavior or mental function. (1976, p. 246)

The key point, for both McCarthy and Rozin, is that it is the specialist demons' accessibility *to each other* (and not to some imagined higher Executive or Central Ego) that could in principle explain the dramatic increases in cognitive competence that we associate with consciousness: the availability to deliberate reflection, the nonautomaticity, in short, the open-mindedness that permits a conscious agent to consider anything in its purview in any way it chooses. This idea was also central to what I called the Multiple Drafts Model (Dennett 1991), which was offered as an alternative to the traditional, and still popular, Cartesian Theater model, which supposes there is a place in the brain to which all the unconscious modules send their results for ultimate conscious appreciation by the Audience. The Multiple Drafts Model did not provide, however, a sufficiently vivid and imagination-friendly antidote to the Cartesian imagery we have all grown up with, so more recently I have proposed what I consider to be a more useful guiding metaphor: "fame in the brain" or "cerebral celebrity" (Dennett 1994a, 1996a, 1998a).

2 Competition for Clout

The basic idea is that consciousness is more like fame than television; it is *not* a special "medium of representation" in the brain into which content-bearing events must be transduced in order to become conscious. As Kanwisher (2001) aptly emphasizes: "the neural correlates of awareness of a given perceptual attribute are found in the very neural structure that perceptually analyzes that attribute." Instead of switching media or going somewhere in order to become conscious, heretofore unconscious contents, staying right where they are, can achieve something *rather like* fame in competition with other fame-seeking (or just potentially fame-*finding*) contents. And, according to this view, that is what consciousness is.

Of course consciousness couldn't be *fame,* exactly, in the brain, since to be famous is to be a shared intentional object *in the conscious minds* of many folk, and although the brain is usefully seen as composed of hordes of demons (or *homunculi*), if we were to imagine them to be *au courant* in the ways they would need to be to elevate some of their brethren to cerebral celebrity, we would be endowing these subhuman components with too much human psychology—and, of course, installing a patent infinite regress in the model as a theory of consciousness. The looming infinite regress can be stopped the way such threats are often happily stopped, not by abandoning the basic idea but by softening it. As long as your *homunculi* are more stupid and ignorant than the intelligent agent they compose, the nesting of homunculi within homunculi can be finite, bottoming out, eventually, with agents so unimpressive that they can be replaced by machines (Dennett 1978).

So consciousness is not so much fame, then, as political influence—a good slang term is *clout.* When processes compete for ongoing control of the body, the one with the greatest clout dominates the scene until a process with even greater clout displaces it. In some oligarchies, perhaps, the only way to have clout is to be *known by the King,* dispenser of all powers and privileges. Our brains are more democratic, indeed somewhat anarchic. In the brain there is no King, no Official Viewer of the State Television Program, no Cartesian Theater, but there are still plenty of quite sharp differences in political clout exercised by contents over time. In Dehaene and Naccache's (2001) terms, this political difference is achieved by "reverberation" in a "sustained amplification loop," while the losing competitors soon fade into oblivion, unable to recruit enough specialist attention to achieve *self-sustaining* reverberation.

What a theory of consciousness needs to explain is how some relatively few contents become elevated to this political power, with all the ensuing *aftermath,* while most others evaporate into oblivion after doing their modest deeds in the ongoing projects of the brain. Why is this the task of a theory of consciousness? Because that is what conscious events do. They hang around, monopolizing time "in the limelight." We cannot settle for putting it that way, however. There is no literal searchlight of attention, so we need to explain away this seductive metaphor by explaining the functional powers of attention-*grabbing* without presupposing a single attention-*giving* source. This means we need to address two questions. Not just (1) How is this fame in the brain achieved? but also (2) And Then What Happens?—which I have called the Hard Question (Dennett 1991, p. 255). One may postulate activity in one neural structure or another as the necessary and sufficient condition for consciousness, but one must then take on the burden of explaining why *that* activity ensures the political power of the events it involves—and this means taking a good hard look at how the relevant differences in competence might be enabled by changes in status in the brain.

Hurley (1998) makes a persuasive case for taking the Hard Question seriously in somewhat different terms: the Self (and its surrogates, the Cartesian *res cogitans*, the Kantian transcendental ego, among others) is not to be located by subtraction, by peeling off the various layers of perceptual and motor "interface" between Self and World. We must reject the traditional "sandwich" in which the Self is isolated from the outside world by layers of "input" and "output." On the contrary, the Self is large, concrete, and visible in the world, not just "distributed" in the brain but spread out into the world. Where we act and where we perceive is not funneled through a bottleneck, physical or metaphysical, in spite of the utility of such notions as *"point of view."* As she notes, the very content of perception can change, while keeping input constant, by changes in output (p. 289).

This interpenetration of effects and contents can be fruitfully studied, and several avenues for future research are opened up by papers in [the special issue of *Cognition* devoted to the cognitive neuroscience of consciousness (in which this chapter appears as an overview)]. What particularly impresses me about them is that the authors are all, in their various ways, more alert to the obligation to address the Hard Question than many previous theorists have been, and the result is a clearer, better focused picture of consciousness in the brain, with no leftover ghosts lurking. If we set aside our *philosophical* doubts (settled or not) about consciousness as global fame or clout, we can explore in a relatively undistorted way the empirical questions regarding the mechanisms and pathways that are necessary, or just normal, for achieving this interesting functional status (we can call it a *Type-C* status, following Jack and Shallice 2001, if we want to remind ourselves of what we are setting aside, while remaining noncommittal).

For example, Parvizi and Damasio (2001) claim that a midbrain panel of specialist proto–self evaluators accomplish a normal, but not necessary, evaluation process that amounts to a sort of triage, which can boost a content into reverberant fame or consign it to oblivion; these proto–self evaluators thereby tend to secure fame for those contents that are most relevant to current needs of the body. Driver and Vuilleumier (2001) concentrate on the "fate of extinguished stimuli," exploring some of the ways that multiple competitions—for example, as proposed by Desimone and Duncan's (1995) Winner-Take-All model of multiple competition—leave not only single winners, but lots of quite powerful semifinalists or also-rans, whose influences can be traced even when they don't achieve the canonical—indeed, operationalized—badge of fame: subsequent reportability (more on that below). Kanwisher (2001) points out that sheer "activation strength" is no mark of consciousness until we see to what use that strength is put ("And then what happens?") and proposes that "the neural correlates of the *contents* of visual awareness are represented in the ventral pathway, whereas the neural correlates of more general-purpose *content-independent* processes associated with awareness (attention, binding, etc.) are found primarily in the dorsal pathway" (p. 98), which suggests (if I understand her claim rightly) that, just as in the wider world, whether or not you become famous can depend on what is going on *elsewhere* at the same time.

Jack and Shallice (2001) propose a complementary balance between prefrontal cortex and anterior cingulate, a sort of high-road versus low-road dual path, with particular attention to the Hard Question: what can happen, what must happen, what may happen when Type-C processes occur, or put otherwise: what Type-C-processes are necessary for, normal for, not necessary for. Particularly important are the ways in which successive winners dramatically alter the prospects (for fame, for influence) of their successors, creating nonce-structures that temporarily govern the competition. Such effects, described at the level of competition between "informations," can begin to explain how one (one agent, one subject) can "sculpt the response space" (Frith 2000; discussed in Jack and Shallice 2001). This downstream capacity of one information to change the competitive context for whatever informations succeed it is indeed a famelike competence, a hugely heightened influence that not only retrospectively distinguishes it from its competitors at the time but also, just as importantly, contributes to the creation of a relatively long-lasting Executive, not a place in the brain but a sort of political coalition that can be seen to *be in control* over the subsequent competitions for some period of time. Such differences in aftermath can be striking, perhaps never more so than those recently demonstrated effects that show, as Dehaene and Naccache (2001) note, "the impossibility for subjects [i.e., Executives] to strategically use the unconscious information," in such examples as Debner and Jacoby 1994 and Smith and Merikle 1999, discussed in Merikle et al. 2001.

Consciousness, like fame, is not an *intrinsic* property, and not even just a *dispositional* property; it is a phenomenon that requires some actualization of the potential—and this is why you cannot make any progress on it until you address the Hard Question and look at the aftermath. Consider the following tale. Jim has written a remarkable first novel that has been enthusiastically read by some of the *cognoscenti*. His picture is all set to go on the cover of *Time Magazine,* and Oprah has lined him up for her television show. A national book tour is planned and Hollywood has already expressed interest in his book. That's all true on Tuesday. Wednesday morning San Francisco is destroyed in an earthquake, and the world's attention can hold nothing else for a month. Is Jim famous? He would have been, if it weren't for that darn earthquake. Maybe next month, if things return to normal, he'll *become* famous for deeds done earlier. But fame eluded him this week, in spite of the fact that the *Time Magazine* cover story had been typeset and sent to the printer, to be yanked at the last moment, and in spite of the fact that his name was already in *TV Guide* as Oprah's guest, and in spite of the fact that stacks of his novels could be found in the windows of most bookstores. All the *dispositional properties* normally sufficient for fame were in place, but their normal effects didn't get triggered, so no fame resulted. The same (I have held) is true of consciousness. The idea of some information being conscious for a few milliseconds, with none of the normal aftermath, is as covertly incoherent as the idea of somebody being famous for a few minutes, with none of the normal aftermath. Jim was potentially famous but didn't quite achieve fame; and he certainly didn't have any *other* property (an eerie glow, an aura of charisma, a threefold increase in "animal magnetism" or whatever) that distinguished him from the equally anonymous people around him. Real fame is not the *cause* of all the normal aftermath; it *is* the normal aftermath.

The same point needs to be appreciated about consciousness, for this is where theorists' imaginations are often led astray: it is a mistake to go looking for an *extra* will-of-the-wisp property of consciousness that might be enjoyed by some events in the brain in spite of their not enjoying the fruits of fame in the brain. Just such a quest is attempted by Block (2001), who attempts to isolate "phenomenality" as something distinct from fame ("global accessibility") but still worthy of being called a variety of consciousness. "Phenomenality is experience," he announces, but what does this mean? He recognizes that in order to keep phenomenality distinct from global accessibility, he needs to postulate, and find evidence for, what he calls "phenomenality without reflexivity"—experiences that you don't know you're having.

If we want to use brain imaging to find the neural correlates of phenomenality, we have to pin down the phenomenal side of the equation and to do that we must make a decision on whether the subjects who say they don't see anything do or do not have phenomenal experiences.

But what then is left of the claim that phenomenality is experience? What is *experiential* (as contrasted with what?) about a discrimination that is not globally accessible? As the convolutions of Block's odyssey reveal, there is always the simpler hypothesis to fend off: there is *potential* fame in the brain (analogous to the dispositional status of poor Jim, the novelist) and then there is fame in the brain, and these two categories suffice to handle the variety of phenomena we encounter. Fame in the brain is enough.

3 Is There Also a Hard Problem?

The most natural reaction in the world to this proposal is frank incredulity: it *seems* to be leaving out the most important element—the Subject! People are inclined to object: "There may indeed be fierce competition between 'informations' for political clout in the brain, but you have left out the First Person, who entertains the winners." The mistake behind this misbegotten objection is not noticing that the First Person has in fact already been incorporated into the multifarious further effects of all the political influence achievable in the competitions. Some theorists in the past have encouraged this mistake by simply stopping short of addressing the Hard Question. Damasio (1999) has addressed our two questions in terms of two intimately related problems: how the brain "generates the movie in the brain" and how the brain generates "the *appearance* of an owner and observer for the movie *within the movie*," and has noted that some theorists, notably Penrose (1989) and Crick (1994), have made the tactical error of concentrating almost exclusively on the first of these problems, postponing the second problem indefinitely. Oddly enough, this tactic is reassuring to some observers, who are relieved to see that these models are not, apparently, denying the existence of the Subject but are just not *yet* tackling that mystery. Better to postpone than to deny, it seems.

A model that, on the contrary, undertakes from the outset to address the Hard Question, assumes the obligation of accounting for the Subject in terms of "a collective dynamic phenomenon that does not require any supervision," as Dehaene and Naccache (2001) put it. This risks seeming to leave out the Subject, precisely because all the work the Subject would presumably have done, once it had enjoyed the show, has already been parceled out to various agencies in the brain, leaving the Subject with nothing to do. We haven't really solved the problem of consciousness until that Executive is itself broken down into subcomponents that are themselves *clearly* just unconscious underlaborers who work (compete, interfere, dawdle, . . .) without supervision. Contrary to appearances, then, those who work on answers to the Hard Question are not leaving consciousness *out*, they are explaining consciousness by leaving it *behind*. That is to say, the only way to explain consciousness is to move beyond consciousness, accounting for the effects consciousness has when it is achieved. It is hard to avoid the nagging feeling, however, that there must be something that such an approach leaves out, something that lies somehow *in between* the causes of consciousness and its effects.

Your body is made up of some trillions of cells, each one utterly ignorant of all the things *you* know. If we are to explain the conscious Subject, one way or another the transition from clueless cells to knowing organizations of cells must be made without any magic ingredients. This requirement presents theorists with what some see as a nasty dilemma (e.g., Brook 2000). If you propose a theory of the knowing Subject that describes whatever it describes as like the workings of a vacant automated factory—not a Subject in sight—you will seem to many observers to have changed the subject or missed the point. On the other hand, if your theory still has tasks for a Subject to perform, still has a need for the Subject as Witness, then although you can be falsely comforted by the sense that there is still somebody at home in the brain, you have actually postponed the task of explaining what needs explaining. To me one of the most fascinating bifurcations in the intellectual world today is between those to whom it is obvious—*obvious*—that a theory that leaves out the Subject is thereby disqualified as a theory of consciousness (in Chalmers's terms, it evades the Hard Problem), and those to whom it is just as obvious that any theory that *doesn't* leave out the Subject is disqualified. I submit that the former have to be wrong, but they certainly don't lack for conviction, as these recent declarations eloquently attest:

> If, in short, there is a community of computers living in my head, there had also better be somebody who is in charge; and, by God, it had better be me. (Fodor 1998, p. 207)

> Of course the problem here is with the claim that consciousness is "identical" to physical brain states. The more Dennett et al. try to explain to me what they mean by this, the more convinced I become that what they really mean is that consciousness doesn't exist. (Wright 2000, ch. 21, n. 14)

> *Daniel Dennett is the Devil.* . . . There is no internal witness, no central recognizer of meaning, and no self other than an abstract "Center of Narrative Gravity" which is itself nothing but a convenient fiction. . . . For Dennett, it is not a case of the Emperor having no clothes. It is rather that the clothes have no Emperor. (Voorhees 2000, pp. 55–56)

This is not just my problem; it confronts anybody attempting to construct and defend a properly naturalistic, materialistic theory of consciousness. Damasio is one who has attempted to solve this pedagogical (or perhaps diplomatic) problem by appearing to split the difference, writing eloquently about the Self, proclaiming that he is taking the Subject very seriously, even *restoring* the Subject to its rightful place in the theory of consciousness—while quietly dismantling the Self, breaking it into "proto-selves" and identifying these in functional, neuroanatomic terms as a network of brain-stem nuclei (Parvizi and Damasio 2001). This effort at winsome redescription, which I applaud, includes some artfully couched phrases that might easily be misread, however, as conceding too much to those who fear that the Subject is being overlooked. One passage in particular goes to the heart of current controversy. They disparage an earlier account that "dates from a time in which the phenomena of consciousness were conceptualized in exclusively behavioral, third-person terms. Little consideration was given to the cognitive, first-person description of the phenomena, that is, to the experience of the subject who is conscious" (p. 136). Notice that they do *not* say that they are now adopting a first-person perspective; they say that they are now giving more consideration to the "first-person *description*" that subjects give. In fact, they are strictly adhering to the canons and assumptions of what I have called *heterophenomenology*, which is specifically designed to be a *third-person* approach to consciousness

(Dennett 1991, ch. 4, "A Method for Phenomenology," p. 98). How does one take subjectivity seriously from a third-person perspective? By taking the *reports* of subjects seriously as reports of their subjective experience. This practice does not limit us to the study of human subjectivity; as numerous authors have noted, nonverbal animals can be put into circumstances in which some of their behavior can be interpreted, as Weiskrantz (1998) has put it, as "commentaries," and Kanwisher (2001) points out that in Newsome's experiments, for instance, the monkey's behavior is "a reasonable proxy for such a report."

It has always been good practice for scientists to put themselves in their own experimental apparatus as informal subjects, to confirm their hunches about what it feels like, and to check for any overlooked or underestimated features of the circumstances that could interfere with their interpretations of their experiments. (Kanwisher gives a fine example of this, inviting the reader into the role of the subject in rapid serial visual display [RSVP], and noting from the inside, as it were, the strangeness of the forced choice task: you find yourself thinking that "tiger" would be as good a word as any, etc.) But scientists have always recognized the need to confirm the insights they have gained from self-administered pilot studies by conducting properly controlled experiments with naive subjects. As long as this obligation is met, whatever insights one may garner from "first-person" investigations fall happily into place in "third-person" heterophenomenology. Purported discoveries that cannot meet this obligation may inspire, guide, motivate, illuminate one's scientific theory, but *they* are not data—the beliefs of subjects about them are the data. Thus if some phenomenologist becomes convinced by her own (first-)personal experience—however encountered, transformed, reflected upon—of the existence of a feature of consciousness in need of explanation and accommodation within her theory, her conviction that this is so is itself a fine datum in need of explanation, by her or by others, but the truth of her conviction must not be presupposed by science. There is no such thing as first-person science, so if you want to have a *science* of consciousness, it will have to be a third-person science of consciousness, and none the worse for it, as the many results discussed in [the special volume of *Cognition*] show.

Since there has been wholesale misreading of this moral in the controversies raging about the "first-person point of view," let me take this opportunity to point out that every study reported in every article [in that special issue of *Cognition*] has been conducted according to the tenets of heterophenomenology. Are the researchers represented here needlessly tying their own hands? Are there other, deeper ways of studying consciousness scientifically? This has recently been claimed by Petitot et al. (1999), who envision a "naturalized phenomenology" that somehow goes beyond heterophenomenology and derives something from a first-person point of view that cannot be incorporated in the manner followed here; but although their anthology includes some very interesting work, it is not clear that any of it finds a mode of scientific investigation that in any way even purports to transcend this third-person obligation. The one essay that makes such a claim specifically, Thompson, Noë, and Pessoa's essay on perceptual completion or "filling in" (see also Pessoa, Thompson, and Noë 1998), corrects some errors in my heterophenomenological treatment of the same phenomena, but it is itself a worthy piece of heterophenomenology, in spite of the authors declarations to the contrary (see Dennett 1998b, and their reply, same issue). Chalmers has made the same unsupported claim:

> I also take it that first-person data can't be expressed wholly in terms of third-person data about brain processes *and the like*. [my italics] . . . That's to say, no purely third-person description of brain processes *and behavior* [my italics] will express precisely the data we want to explain, though it may play a central role in the explanation. So "as data," the first-person data are irreducible to third-person data. (1999, p. 8)

This swift passage manages to overlook the prospects of heterophenomenology altogether. Heterophenomenology is explicitly not a first-person methodology (as its name makes clear) but it is also not directly about "brain processes and the like"; it is a reasoned, objective extrapolation from patterns discernible in the behavior of subjects, including especially their text-producing or communicative behavior, and as such it is *about* precisely the higher-level dispositions, both cognitive and emotional, that convince us that our fellow human beings are conscious. By sliding from the first italicized phrase to the second (in the quotation above), Chalmers executes a (perhaps unintended) sleight-of-hand, whisking heterophenomenology off the stage without a hearing. His conclusion is a non sequitur. He has not shown that first-person data are irreducible to third-person data because he has not even considered the only serious attempt to show *how* first-person data can be "reduced" to third-person data (though I wouldn't use that term).

The third-person approach is not antithetical to, or eager to ignore, the subjective nuances of experience; it simply insists on anchoring those subjective nuances to *something*—anything, really—that can be detected and confirmed in replicable experiments. For instance, Merikle et al. (2001), having adopted the position that "with subjective measures, awareness is assessed on the basis of the observer's self-reports," note that one of the assumptions of this approach is that "information perceived with awareness enables a perceiver to act on the world and to produce effects on the world." As contrasted to what? As contrasted to a view, such as that of Searle (1992) and Chalmers (1996), that concludes that consciousness *might* have no such enabling role—since a "zombie" might be able to do everything a conscious person does, passing every test, reporting every effect, without being conscious. One of the inescapable implications of heterophenomenology, or of any third-person approach to subjectivity, is that one must dismiss as a chimera the prospect of a philosopher's zombie, a being that is behaviorally, objectively indistinguishable from a conscious person but not conscious. (For a survey of this unfortunate topic, see *Journal of Consciousness Studies* 2, 1995, "Zombie Earth: A Symposium," including short pieces by many authors.)

I find that some people are cured of their attraction for this chimera by the observation that all the functional distinctions described in the essays in [the special volume of *Cognition*] would be exhibited by philosophers' zombies. The only difference between zombies and regular folks, according to those who take the distinction seriously, is that zombies have streams of *unconsciousness* where the normals have streams of consciousness! Consider, in this regard, the word-stem completion task of Debner and Jacoby (1994) discussed by Merikle et al. (2001). If subjects are instructed to complete a word stem with a word other than the word briefly presented as a prime (and then masked), they can follow this instruction only if they are aware of the priming word; they actually favor the priming word as a completion if it is presented so briefly that they are not aware of it. Zombies would exhibit the same effect, of course—being able to follow the exclusion policy only in those instances in which the priming word made it through the competition into their streams of *un*consciousness.

4 But What about "Qualia"?

As Dehaene and Naccache (2001) note:

> [T]he flux of neuronal workspace states associated with a perceptual experience is vastly beyond accurate verbal description or long-term memory storage. Furthermore, although the major organization of this repertoire is shared by all members of the species, its details result from a developmental process of epigenesis and are therefore specific to each individual. Thus the contents of perceptual awareness are complex, dynamic, multi-faceted neural states that cannot be memorized or transmitted to others in their entirety. These biological properties seem potentially capable of substantiating philosophers' intuitions about the "qualia" of conscious experience, although considerable neuroscientific research will be needed before they are thoroughly understood.

It is this informational superabundance, also noted by Damasio (1999, see esp. p. 93), that has lured philosophers into a definitional trap. As one sets out to answer the Hard Question ("And then what happens?"), one can be sure that no practical, finite set of answers will exhaust the richness of effects and potential effects. The subtle individual differences wrought by epigenesis and a thousand chance encounters create a unique manifold of functional (including *dys*functional) dispositions that outruns any short catalog of effects. These dispositions may be dramatic—ever since that yellow car crashed into her, one shade of yellow sets off her neuromodulator alarm floods (Dennett 1991)—or minuscule—an ever so slight relaxation evoked by a nostalgic whiff of childhood comfort food.

So one will always be "leaving something out." If one dubs this inevitable residue *qualia,* then qualia are guaranteed to exist, but they are just more of the same, dispositional properties that have not yet been entered in the catalog (perhaps because they are the most subtle, least amenable to approximate definition). Alternatively, if one defines *qualia* as whatever is neither the downstream effects of experiences (reactions to particular colors, verbal reports, effects on memory, etc.) nor the upstream causal progenitors of experiences (activity in one cortical region or another), then qualia are, by definitional fiat, *intrinsic properties* of experiences considered in isolation from all their causes and effects, logically independent of all dispositional properties. Defined thus, they are logically guaranteed to elude all broad functional analysis—but it's an empty victory, since there is no reason to believe such properties exist!

To see this point more clearly, compare the qualia of experience to the value of money. Some naive Americans cannot get it out of their heads that dollars, unlike francs and marks and yen, have *intrinsic value* ("How much is that in *real* money?"). They are quite content to "reduce" the value of other currencies in dispositional terms to their exchange rate with dollars (or goods and services), but they have a hunch that dollars are different. Every dollar, they declare, has something logically independent of its functionalistic exchange powers, which we might call its *vim.* So defined, the *vim* of each dollar is guaranteed to elude the theories of economists forever, but we have no reason to believe in it—aside from their heartfelt hunches, which can be explained without being honored. It is just such an account of philosophers' intuitions that Dehaene and Naccache propose.

It is unfortunate that the term, *qualia*, has been adopted—in spite of my warnings (1988, 1991, 1994c)—by some cognitive neuroscientists who have been unwilling or unable to believe that philosophers intend that term to occupy a peculiar logical role in arguments about functionalism that cognitive neuroscience *could not* resolve. A review of recent history will perhaps clarify this source of confusion and return us to the real issues. (This next passage repeats some material from chapter 1.)

Functionalism is the idea enshrined in the old proverb: handsome is as handsome does. Matter matters only because of what matter can do. Functionalism in this broadest sense is so ubiquitous in science that it is tantamount to a reigning presumption of all of science. And since science is always looking for simplifications, looking for the greatest generality it can muster, functionalism in practice has a bias in favor of minimalism, of saying that less matters than one might have thought. The law of gravity says that it doesn't matter what stuff a thing is made of—only its mass matters (and its density, except in a vacuum). The trajectory of cannonballs of equal mass and density is not affected by whether they are made of iron, copper, or gold. It might have mattered, one imagines, but in fact it doesn't. And wings don't have to have feathers on them in order to power flight, and eyes don't have to be blue or brown in order to see. Every eye has many more properties than are needed for sight, and it is science's job to find the maximally general, maximally noncommittal—hence minimal—characterization of whatever power or capacity is under consideration. Not surprisingly, then, many of the disputes in normal science concern the issue of whether or not one school of thought has reached too far in its quest for generality.

Since the earliest days of cognitive science, there has been a particularly bold brand of functionalistic minimalism in contention: the idea that just as a heart is basically a pump, and could in principle be made of anything so long as it did the requisite pumping without damaging the blood, so a mind is fundamentally a control system, implemented in fact by the organic brain, but anything else that could *compute the same control functions* would serve as well. The actual matter of the brain—the chemistry of synapses, the role of calcium in the depolarization of nerve fibers, and so forth—is roughly as irrelevant as the chemical composition of those cannonballs. According to this tempting proposal, even the underlying microarchitecture of the brain's connections can be ignored for many purposes, at least for the time being, since it has been proven by computer scientists that any function that can be computed by one specific computational architecture can also be computed (perhaps much less efficiently) by another architecture. If all that matters is the computation, we can ignore the brain's wiring diagram, and its chemistry, and just worry about the "software" that runs on it. In short—and now we arrive at the provocative version that has caused so much misunderstanding—in principle you could replace your wet, organic brain with a bunch of silicon chips and wires and go right on thinking (and being conscious, and so forth).

This bold vision, computationalism or "strong AI" (Searle 1980), is composed of two parts: the broad creed of functionalism—handsome is as handsome does—and a specific set of minimalist empirical wagers: neuroanatomy doesn't matter; chemistry doesn't matter. This second theme excused many would-be cognitive scientists from educating themselves in these fields, for the same reason that economists are excused from knowing anything about the metallurgy of coinage, or the chemistry of the ink and paper used in bills of sale. This has been a good idea in many ways, but for fairly obvious reasons, it has not been a politically astute ideology, since it has threatened to relegate those scientists who devote their lives to functional neuroanatomy and neurochemistry, for instance, to relatively minor roles as electricians and plumbers in the grand project of explaining consciousness. Resenting this proposed demotion, they have fought back vigorously. The recent history of neuroscience can be seen as a series of triumphs for the lovers of detail. Yes, the specific geometry of the connectivity matters; yes, the location of specific neuromodulators and their effects matter; yes, the architecture matters; yes, the fine temporal rhythms of the spiking patterns matter, and so on. Many of the fond hopes of opportunistic minimalists have been dashed: they had hoped they could leave out various things, and they have learned that no, if you leave out x, or y, or z, you can't explain how the mind works.

This result has left the mistaken impression in some quarters that the underlying idea of functionalism has been taking its lumps. Far from it. On the contrary, the reasons for accepting these new claims are precisely the reasons of functionalism. Neurochemistry matters because—and *only* because—we have discovered that the many different neuromodulators and other chemical messengers that diffuse through the brain have *functional roles* that make important differences. What those molecules *do* turns out to be important to the *computational* roles played by the neurons, so we have to pay attention to them after all.

This correction of overly optimistic minimalism has nothing to do with philosophers' imagined *qualia*. Some neuroscientists have thus muddied the waters by befriending qualia, confident that this was a term for the sort of functionally characterizable complication that confounds oversimplified versions of computationalism. (Others have thought that when philosophers were comparing zombies with conscious people, they were noting the importance of emotional state, or neuromodulator imbalance.) I have spent more time that I would like explaining to various scientists that their controversies and the philosophers' controversies are not translations of each other as they had thought but false friends, mutually irrelevant to each other. The principle of charity continues to bedevil this issue, however, and many scientists generously persist in refusing to believe that philosophers can be making a fuss about such a narrow and fantastical division of opinion.

Meanwhile, some philosophers have misappropriated those same controversies within cognitive science to support their claim that the tide is turning against functionalism, in favor of qualia, in favor of the irreducibility of the "first-person point of view," and so forth. This widespread conviction is an artifact of interdisciplinary miscommunication and nothing else. A particularly vivid exposure of the miscommunication can be found in the critics' discussion of Humphrey 2000. In his rejoinder, Humphrey says:

> I took it for granted that everyone would recognise that my account of sensations was indeed meant to be a functional one through and through—so much so that I actually deleted the following sentences from an earlier draft of the paper, believing them redundant: "Thus [with this account] we are well on our way to doing the very thing it *seemed* we would not be able to do, namely giving the mind term of the identity, the phantasm, a *functional description*—even if a rather unexpected and peculiar one. And, as we have already seen, once we

have a functional description we're home and dry, because the same description can quite well fit a brain state."

But perhaps I should not be amazed. Functionalism is a wonderfully—even absurdly—bold hypothesis, about which few of us are entirely comfortable.

5 Conclusion

A neuroscientific theory of consciousness must be a theory of the Subject of consciousness, one that analyzes this imagined central Executive into component parts, none of which can itself be a proper Subject. The apparent properties of consciousness that make sense only as *features enjoyed by the Subject* must thus also be decomposed and distributed, and this inevitably creates a pressure on the imagination of the theorist. No sooner do such properties get functionalistically analyzed into complex dispositional traits distributed in space and time in the brain, than their ghosts come knocking on the door, demanding entrance disguised as *qualia*, or *phenomenality*, or *the imaginable difference between us and zombies*. One of the hardest tasks thus facing those who would explain consciousness is recognizing when some feature has *already* been explained (in sketch, in outline) and hence does not need to be explained again.

Many years ago, a friend told me about a professor of literature who was puzzled by a final examination essay in which a student went on at some length about *fantasy echo* poetry.[1] The professor called the student in and queried him about his curiously evocative but unexplained epithet. What on earth was the student talking about and where, if one might ask, had he picked up this idea? "From your lectures, of course!" the student replied. The professor was dumfounded, but soon enough got to the bottom of the mystery: he had often referred in his lectures to late-nineteenth-century works in the *fin de siècle* style.

This gem of serendipitous misperception has been rattling around in my brain for several decades. A few months ago, it occurred to me that it really deserved a new career, and that the time was ripe for reincarnation. Eureka! For those same decades I had been yearning for a sailboat; this was the year, at last, to

1. This chapter is drawn with many deletions and revisions from a lecture I gave at the Consciousness in London Conference at The Kings College, London, April 24, 1999. That lecture included many of the passages included in earlier chapters, and these have largely been excised, leaving behind just a few expressions and arguments that may help clarify my view.

buy a boat, and its name shall be *Fantasy Echo*. What a perfect name for a 1999 dreamboat! But for various good reasons it turns out that 1999 is *not* a good year for me to buy a sailboat (chartering, once again, must suffice); it appears that my relief from slooplessness, as Quine once put it, will have to await the next millennium. What my university's fund-raisers call a *great naming opportunity* was going to slip away from me, unexploited. What a pity!

I was recounting all this to another friend recently, who startled me by pointing out that I already owned, and had been working on for years, something that could with even more justice be named *Fantasy Echo*: my theory of human consciousness. So with a little help from my friends, I am happy to unveil, at this 1999 conference on theories of consciousness, my updated and newly renamed *Fantasy Echo* theory of consciousness.

1   Fleeting Fame

This is the theory that went by the name of the Multiple Drafts Model in 1991, and has more recently been advertised by me as the "fame in the brain" (or "cerebral celebrity") model (1996b, 1998a, 2001a). The basic idea is that consciousness is more like fame than television; it is *not* a special "medium of representation" in the brain into which content-bearing events must be "transduced" in order to become conscious. It is rather a matter of content-bearing events in the brain achieving something a bit like fame in competition with other fame-seeking (or at any rate potentially fame-finding) events.

But of course consciousness couldn't be *fame*, exactly, in the brain, since to be famous is to be a shared intentional object *in the consciousnesses* of many folk, and although the brain is usefully seen as composed of hordes of *homunculi*, imagining them to be *au courant* in just the way they would need to be to elevate some of their brethren to cerebral celebrity is going a bit too far—to say nothing of the problem that it would install a patent infinite regress in my theory of consciousness. The looming infinite regress can be stopped the way such threats are often happily stopped, not by abandoning the basic idea but by softening it. As long as your *homunculi* are more stupid and ignorant than the intelligent agent they compose, the nesting of homunculi within homunculi can be finite, bottoming out, eventually, with agents so unimpressive that they can be replaced by machines.

So consciousness is not so much *fame*, then, as *influence*—a species of relative "political" power in the opponent processes that eventuate in ongoing control of the body. In some oligarchies, perhaps, the only way to achieve political power is to be *known by the King*, dispenser of all powers and privileges. Our brains are more democratic, indeed anarchic. In the brain there is no King, no Official Viewer of the State Television Program, no Cartesian Theater, but there are still plenty of *quite* sharp differences in political power exercised by contents over time. What a theory of consciousness needs to explain is how some relatively few contents become elevated to this political power, while most others evaporate into oblivion after doing their modest deeds in the ongoing projects of the brain.

Why is *this* the task of a theory of consciousness? Because that is what conscious events *do*. They hang around, monopolizing time "in the limelight"—but we need to explain *away* this seductive metaphor, and its kin, the searchlight of attention, by explaining the *functional* powers of attention-*grabbing* without presupposing a single attention-*giving* source. That is the point of what I call the Hard Question: And Then What Happens? Postulate activity in whatever neural structures you please as the necessary and sufficient condition for consciousness, but then take on the burden of explaining why *that* activity ensures the political power of the events it involves.

The attractiveness of the idea of a special medium of consciousness is not simply a persistent hallucination. It is not *entirely* forlorn, as we can see by pursuing the analogy with fame a bit further. Fame—in the world, not in the brain—is not what it used to be. The advent of new media of communication has in fact radically changed the nature of fame, and of political power, in our social world, and something interestingly analogous may have happened in the brain. That, in any case, is my speculative proposal. As I have argued over and over again, *being in consciousness* is not like *being on television*; one can be on television and be seen by millions of viewers, and still not be famous, because one's television debut does not have the proper *sequelae*. Similarly, there is no special area in the brain where representation is, by itself, sufficient for consciousness. It is always the *sequelae* that make the difference. (And Then What Happens?)

My inspiration for the fame-in-the-brain analogy was, of course, Andy Warhol:

In the future, everybody will be famous for fifteen minutes.

What Warhol nicely captured in this remark was a reductio ad absurdum of a certain (imaginary) concept of fame. Would that be *fame*? Has Warhol described a logically possible world? If we pause to think about it more carefully than usual, we see that something has been stretched beyond the breaking point. It is true, no doubt, that thanks to the mass media, fame can be conferred on an anonymous citizen almost instantaneously (Rodney King comes to mind), and thanks to the fickleness of public attention, can evaporate almost as fast, but Warhol's rhetorical exaggeration of this fact carries us into the absurdity of Wonderland. We have yet to see an instance of someone being famous for just fifteen minutes, and in fact we never will. Let some citizen be viewed for fifteen minutes or less by hundreds of millions of people, and then—unlike Rodney King—be utterly forgotten. To call that fame would be to misuse the term (ah yes, an "ordinary language" move, and a good one, if used with discretion). If that is not obvious, then let me raise the ante: could a person be famous *for five seconds* (not merely attended-to-by-millions of eyes but famous)? There are in fact hundreds if not thousands of people who every day pass through the state of being viewed, for a few seconds, by millions of people. Consider the evening news, presenting a story about the approval of a new drug. An utterly anonymous doctor is seen (by millions) plunging a hypodermic into the arm of an utterly anonymous patient—that's being on television, but it isn't fame!

Several philosophers have risen to the bait of my rhetorical question and offered counterexamples to my implied claim about the duration of fame. Here is how somebody could be famous for fifteen seconds: he goes on international TV, introduces himself as the person who is about to destroy our planet and thereupon does so. Oh, they got me! But notice that this example actually works in my favor. It draws attention to the importance of the normal *sequelae*: the only way to be famous for less than a longish time is to destroy the whole world in which your fame would otherwise reverberate. And if anybody wanted to cavil about whether that was *really* fame, we could note how the question could be resolved in an extension of the thought experiment. Suppose our antihero presses the button and darn, no nuclear explosion! *And then what happens?* The world survives, and in it we either observe the normal *sequelae* of fame or we don't. In the latter case, we would conclude, retrospectively, that our candidate's bid for fame had simply failed, in spite of his widely broadcast image. (Maybe nobody was watching, or paying attention.) The important point of the analogy is that consciousness, like fame, is a *functionalistic* phenomenon: handsome is as handsome *does*.

The importance of such echoes, of *reverberation*, of *return-trips*, of *reminding*, of *recollectability*, is often noted by writers on consciousness. Here is Richard Powers, for instance:

> To remember a feeling without being able to bring it back. This seemed to me as close to a functional definition of higher-order consciousness as I would be able to give her. (1995, p. 228)

But this is "higher-order" consciousness, isn't it? What about "lower-order" consciousness? Might the echo-capacity be wholly absent therefrom? The idea that we can identify a variety of consciousness that is logically independent of the echo-making power has many expressions in the recent literature. It is even tempting to suppose that this lower or simpler variety of consciousness is somehow a normal precondition for echo-making. It is, perhaps, the very feature that echoes when there *are* echoes. A particularly popular version is Ned Block's proposed distinction between *phenomenal consciousness* and *access consciousness*. Fame in the brain provides, perhaps, a useful way of thinking about the "political" access that some contents may have to the reins of power in the ongoing struggle to control the body, but it has nothing to say about the brute, lower-order, *what-it-is-like-ness* of phenomenal consciousness.

What is it like *to whom*? As I have often said, in criticism of Block's attempted distinction, once you shear off all implications about "access" from phenomenal consciousness, you are left with something apparently indistinguishable from phenomenal *un*consciousness. Consider an example. As a left-handed person, I can wonder whether I am a left-hemisphere-dominant speaker or a right-hemisphere-dominant speaker or something mixed, and the only way I can learn the truth is by submitting myself to objective, "third-person" testing. I don't "have access to" this intimate fact about how my own mind does its work. It escapes all my attempts at introspective detection, and might, for all I know, shunt back and forth every few seconds without my being any the wiser (see chapter 4). This is just one of many—indeed countless—"intrinsic" properties that the events occurring in my brain have that, by being entirely inaccessible to me, are paradigms of unconscious properties. The challenge facing those who want to claim that some among these "intrinsic" properties are the properties of phenomenal *consciousness* is to show what makes them different (without making any appeal to "access" or echo-making power).

It is the echo-making power, after all, that we invariably appeal to when we try to motivate the claims we make about the consciousness not just of others, but of ourselves. Proust famously elevated the really quite delicate aroma of *madeleines*, almond cookies, for its power to provoke in him vivid memories and emotions from his childhood. The inviting aroma of classroom library paste (safe, edible!) has a similar effect on me. Contrast it with the aroma of, say, the Formica desktop at which

I sat in second grade. But, you protest, it doesn't have an aroma! Well, it does, but not an evocative one for me, not one whose coming and going I can even detect under normal circumstances. It is an aroma that is, at best, *subliminal*—beneath the threshold of my consciousness. What if, nevertheless, it could be shown that the presence or absence of that desktop in my olfactory environment had a subtle biasing effect on my performance on some cognitive task—it might, for instance, bias me in favor of thinking first of the most classroom-relevant meanings of ambiguous words. If so, we'd have a quandary: was this, like "blindsight in normals," a case of *un*conscious echo, or a proof that the aroma of the Formica was indeed part of the background (the Background, to some) of my boyhood consciousness? Either way, it is the presence of an echo, however faint, that provides whatever motivation the latter view has. The believer in phenomenal consciousness stripped even of this echo-making power has a tough sell: the coming and going of the aroma is a change in phenomenal consciousness in spite of the subject's total obliviousness—lack of access—to it.

I said earlier that the idea of consciousness depending on a special medium of representation in the brain is not entirely forlorn, and with these clarifications of our intuitions behind us, I am ready to tackle that issue. Television and fame are two entirely different sorts of things—one's a medium of representation and one isn't—but the sorts of fame made possible by television are interestingly different from earlier sorts of fame, as we have recently been told 'til we're sick of it. Consider the phenomena of Princess Diana, O. J. Simpson, and Monica Lewinsky. In each case a recursive positive feedback became established, dwarfing the initial triggering event and forcing the world to wallow in wallowing in wallowing in reactions to reactions to reactions to the coverage in the media of the coverage in the media of the coverage in the media, and so forth. Did similar fame-phenomena occur in the preelectronic age? The importance of publicity had been appreciated for millennia—secret coronations, for instance, have always been shunned, for the obvious reasons. There have long been sites of recursive reaction, such as the page of letters to the editor in the *Times* (of London, and to a lesser extent, the *New York Times*). But these were still relatively slow, "narrow band" (as we say nowadays) channels of communication, and they reached a small but influential segment of the populace. In the preelectronic age, were there people who were famous for being famous? It is the capacity for the combined modern media to capture *anything* and turn it into a ubiquitously "accessible" or "influential" topic through sheer echoic amplification that strikes some observers as a novel (and perhaps alarming) social phenomenon, and I want to suggest that a similar family of innovations in the brain may lie behind the explosive growth in *reflective power* that I take to be the hallmark of consciousness.

2 Instant Replay

At this point in earlier discussions of this topic, the loyal opposition notes that I am impressed—perhaps overimpressed—with the power of *self*-consciousness, or *reflective* or *introspective* consciousness, at the expense of just plain animal *sentience* or, echoing Block again, *phenomenal* consciousness, but when I talk of reflective power here, I am *not* talking about the highly intellectual (and arguably language-dependent) capacity for—shall we say—*musing* about our *musings*. I'm talking about the capacity of a dog, for instance, to be reminded of its owner or its tormentor by an aroma that provokes an echo that provokes a reidentification. But if *that* is all I'm talking about, then the objection still stands: my notorious claim that human consciousness is largely a culturally borne "meme machine" is refuted by the example of the dog!

Not so fast. It would be refuted—or at least somewhat displaced—by the dog if we could be sure that the reminding aroma really does operate by triggering in the dog the sort of echoic, Proustian events that we report to each other. But there may well be simpler hypotheses that explain the dog's delighted (or hostile) arousal when the aroma hits its nostrils. What else might be going on when a dog "recognizes" somebody by aroma? Does—can—the dog *recollect* the earlier encounter? Are dogs capable of *episodic memory*, or is there just summoned up in the dog a "visceral" echo, of either joy or fear? Minimal recognition of this sort need not involve recollection in our own case, so it need not involve recollection in the case of other species. It need not bring in its ensemble the Proustian trappings and surroundings of the earlier encounter that normally—but not always—decorate our own episodes of episodic memory.

These added details are not just decorations, of course. We human beings rely on them to confirm to ourselves that we are indeed remembering, and not just imagining or guessing. Did I ever meet C. I. Lewis? Yes, once. He was a very old man, and I was a freshman at Wesleyan, in 1959–1960, and he came to give a lecture or two there. I didn't know anything about him at the time, but my philosophy professor had encouraged me to attend, just to see a great man. It was in the Honors College, I recall, and he sat down to read his paper (and I was sitting on the north side of the room facing him, as I recall)—but I don't recall what his paper was about at all. I was more impressed by the respect he was shown by all in attendance than by anything he actually said. The next year, I read two of his books in Roderick Firth's epistemology class at Harvard, which cemented the recollections I'm now reviewing, and of course completely contaminated any memories I might have otherwise thought I had about what he'd said the year before. Now perhaps dogs have similar reflective episodes in their inner lives; if they do, then surely they are just as conscious as we are, in every sense. But I hypothesize—this is the empirical going-out-on-a-limb part of my view—that they do not. Events in their echo chambers damp down to nothing after a couple of reverberations, I suspect. Why? Because they do not need such an echo chamber for anything, and it is not a by-product of anything else they need, and it would be expensive. For nonhuman animals, I suspect, efficiency and timeliness are the desiderata that dictate short, swift, ballistic trajectories of contents. As the business consultants say, the goal is: *Up your throughput!*

But we human beings got sidetracked. We developed a habit of "replaying events in our minds" over and over, and this habit, initially "wasteful" of time and energy, is very likely the source of one of our greatest talents: episodic memory and "one-shot learning" that is not restricted to special cases. (The Garcia Effect is one such special case; rats made nauseous while eating a novel-smelling food have a remarkable Proust effect indeed: they develop an instant distaste for anything with that smell.)

Scientists who use animals in experiments know that in order to teach a new habit, a new discrimination, to an animal, they will typically have to repeat a training or conditioning episode, sometimes three or four times, sometimes hundreds or even thousands of times, before the animal reliably extracts the desired content. There is "one-shot" learning of particularly galvanizing lessons, but can the learner later recall the episode or just the lesson? Might it be that our familiar human talent for reliving pastel versions of our earlier experiences is in large part a learned trick? The events we can readily recall from our lives are actually a rather limited subset of what happens during our waking lives. (Or can you dredge up what you were thinking about while you brushed your teeth last Wednesday?) Episodes in real life happen just once, without (external) repetition, but perhaps our habit of immediately reviewing or rehearsing whatever grabs our attention strongly is a sort of inadvertent self-conditioning that drives these events into the imaginary "storehouse of episodic memory" (it is certainly not an organ or subsystem of the brain). The hypothesis is that until you've acquired the habit of such "instant replay," permitting the choice bits of daily life to reverberate for a while in the brain, you won't have any episodic memory. This could account for "infantile amnesia," of course, and a further, independent hypothesis is that it is a humans-only phenomenon, an artifact of habits of self-stimulation that other species can't acquire in the normal course of things.

Episodic memory is not for free. One idea is that it is the very echoic power that makes episodic memory possible. Animals remember thanks to multiple repetitions of stimuli in the world. We remember, it seems, one-shot, but really, it isn't just one-shot. What we remember is stuff that has been played and replayed and replayed obsessively in our brains. (Note that a feature, not a bug, in this account is that although some repetition is indeed all too familiar to us as *conscious* repetition, the repetition that *elevates* a content to the clout of conscious recallability is largely not conscious. Indeed, there is no need for a sharp dividing line between conscious and unconscious repetitions. No bright line need distinguish true fame from mere behind-the-scenes influence.) It is the echo that creates the capacity for long-term episodic memory. We are used to using these trappings as confirmations of our own convictions that we *are* recollecting. Did you ever meet Carnap? Yes, I reply. "It was at UCLA, in 1965 or '66, I would guess. It was in the corridor outside the philosophy department, and as best I recall, Alfred Tarski and Richard Montague were talking with Carnap. I asked somebody who the people with Tarski were, and when they told me, I just couldn't resist going up and barging in and just shaking their hands."

This instant-reply habit itself has its amusing analogue in the world of electronic media. Before the existence of *videotape*, being on television was not a particularly echoic phenomenon. The programs were broadcast "live" and once they were over, they were over—echoing for awhile in the memories and discussions of the audience, but quick to damp out and slide into oblivion. Newsreels at the cinema were different. Newspapers were different. They preserved for review the events of the day. Until memory was added, radio and television were *not* the sort of media that could provide a suggestive hint about the structure—and media—of consciousness, since their contents were utterly evanescent, no better, really, than the flitting images on the blank wall of the *camera obscura*—except in the memories of those who witnessed them.

Let me sum up. I have ventured (1) the empirical hypothesis that our capacity to relive or rekindle contentful events is the most important feature of consciousness—indeed, as close to a defining feature of consciousness as we will ever find; and (2) the empirical hypothesis that this echoic capacity is due in large part to habits of self-stimulation that we pick up from human culture, that the Joycean machine in our brains is a virtual machine made of memes. These are independent claims. If the meme-hypothesis were roundly defeated by the discovery—the confirmation—of just such echoic systems at play in the brains of nonhuman animals, I would then agree, for that very reason, that the species having those echo-chambers were conscious in just about the way we are—because that's what I say consciousness is. The price I'd pay for that verdict is the defeat of my bold claim about software and *virtual* machines, but I'd still be getting a bargain, since the other side would be relying on the fame theory of consciousness *as a theory of consciousness* in order to establish the relevance—to riddles about consciousness—of their discoveries.

Consciousness often seems to be utterly mysterious. I suspect that the principle cause of this bafflement is a sort of accounting error that is engendered by a familiar series of challenges and responses. A simplified version of one such path to mysteryland runs as follows:

Phil: What is consciousness?

Sy: Well, some things—such as stones and can-openers—are utterly lacking in any *point of view*, any *subjectivity* at all, while other things—such as you and me—do have points of view: private, perspectival, interior ways of being apprised of some limited aspects of the wider world and our bodies' relations to it. We lead our lives, suffering and enjoying, deciding and choosing our actions, guided by this "first-person" access that we have. To be conscious is to be an agent with a point of view.

Phil: But surely there is more to it than that! A cherry tree has limited access to the ambient temperature at its surface, and can be (mis-)guided into blooming inopportunely by unseasonable warm weather; a robot with video camera "eyes" and microphone "ears" may discriminate and respond aptly to hundreds of different aspects of its wider world; my own immune system can sense, discriminate, and respond appropriately (for the most part) to millions of different eventualities. Each of these is an agent (of sorts) with a point of view (of sorts) but none of them is conscious.

Sy: Yes, indeed; there is more. We conscious beings have capabilities these simpler agents lack. We don't just notice things and respond to them; we *notice* that we notice things. More exactly, among the many discriminative states that our bodies may enter (including the states of our immune systems, our autonomic nervous systems, our digestive systems, and so forth), a subset of them can be discriminated in turn by higher-order discriminations which then become sources of guidance for higher-level control activities. In us, this recursive capacity for self-monitoring exhibits no clear limits—beyond those of available time and energy. If somebody throws a brick at you, you see it coming and duck. But you also discriminate the fact that you *visually* discriminated the projectile, and can then discriminate the further fact that you can tell visual from tactile discriminations (usually), and then go on to reflect on the fact that you are also able to recall recent sensory discriminations in some detail, and that there is a difference between experiencing something and recalling the experience of something, and between thinking about the difference between recollection and experience and thinking about the difference between seeing and hearing, and so forth, 'til bedtime.

Phil: But surely there is more to it than that! Although existing robots may have quite paltry provisions for such recursive self-monitoring, I can readily imagine this particular capacity being added to some robot of the future. However deftly it exhibited its capacity to generate and react appropriately to "reflective" analyses of its underlying discriminative states, it wouldn't be conscious—not the way we are.

Sy: Are you sure you can imagine this?

Phil: Oh yes, absolutely sure. There would be, perhaps, some sort of *executive* point of view definable by analysis of the power such a robot would have to control itself based on these reactive capacities, but this robotic subjectivity would be a pale shadow of ours. When it uttered "it seems to me . . . ," its utterances wouldn't really mean anything—or at least, they wouldn't mean what I mean when I tell you what it's like to be me, how things seem to me.

Sy: I don't know how you can be so confident of that, but in any case, you're right that there is more to consciousness than that. Our discriminative states are not just discriminable; they have the power to provoke preferences in us. Given choices between them, we are not indifferent, but these preferences are themselves subtle, variable, and highly dependent on other conditions. There is a time for chocolate and a time for cheese, a time for blue and a time for yellow. In short (and oversimplifying hugely), many if not all of our discriminative states have what might be called a dimension of affective valence. We care which states we are in, and this caring is reflected in our dispositions to change state.

Phil: But surely there is more to it than that! When I contemplate the luscious warmth of the sunlight falling on that old brick wall, it's not just that I prefer looking at the bricks to looking down at the dirty sidewalk beneath them. I can readily imagine outfitting our imaginary robot with built-in preferences for every possible sequence of its internal states, but it would still not have anything like my conscious *appreciation* of the visual poetry of those craggy, rosy bricks.

Sy: Yes, I grant it; there is more. For one thing, you have metapreferences; perhaps you wish you could stop those sexual associations from interfering with your more exalted appreciation of the warmth of that sunlight on the bricks, but at the same time (roughly) you are delighted by the persistence of those saucy intruders, distracting as they are, but . . . what was it you were trying to think about? Your stream of consciousness is replete with an apparently unending supply of associations. As each fleeting occupant of the position of greatest influence gives way to its successors, any attempt to halt this helter-skelter parade and monitor the details of the associations only generates a further flood of evanescent states, and so on. Coalitions of themes and projects may succeed in dominating "attention" for some useful and highly productive period of time, fending off would-be digressions for quite a while, and creating the sense of an abiding self or ego taking charge of the whole operation. And so on.

Phil: But surely there is more to it than that! And now I begin to see what is missing from your deliberately evasive list of additions. All these dispositions and metadispositions to enter into states and metastates and metametastates of reflection about reflection could be engineered (I dimly imagine) into some robot. The trajectory of its internal state-switching could, I suppose, look strikingly similar to the "first-person" account I might give of my own stream of consciousness, but those states of the robot would have no actual *feel*, no *phenomenal* properties at all! You're still leaving out what the philosophers call qualia.

Sy: Actually, I'm still leaving out *lots* of properties. I've hardly begun acknowledging all the oversimplifications of my story so far, but now you seem to want to preempt any further additions from me by insisting that there are properties of consciousness that are altogether different from the properties I've described so far. I thought I *was* adding "phenomenal" properties in response to your challenge, but now you tell me I haven't even begun. Before I can tell if I'm leaving these properties out, I have to know what they are. Can you give me a clear example of a phenomenal property? For instance, if I used to like a particular shade of yellow, but thanks to some traumatic experience (I got struck by a car of that color, let's suppose), that shade of yellow now makes me very uneasy (whether or not it reminds me explicitly of the accident), would this suffice to change the *phenomenal* properties of my experience of that shade of yellow?

Phil: Not necessarily. The *dispositional* property of making you uneasy is not itself a phenomenal property. Phenomenal properties are, by definition, not dispositional but rather intrinsic and accessible only from the first-person point of view . . .

Thus we arrive in mysteryland. If you *define* qualia as *intrinsic properties* of experiences considered in isolation from all their causes and effects, logically independent of all dispositional properties, then they are logically guaranteed to elude all broad functional analysis—but it's an empty victory, since there is no reason to believe such properties exist. To see this, compare the *qualia* of experience to the *value* of money. Some naive Americans can't get it out of their heads that dollars, unlike francs and marks and yen, have *intrinsic value* ("How much is that in *real* money?"). They are quite content to "reduce" the value of other currencies in dispositional terms to their exchange rate with dollars (or goods and services), but they have a hunch that dollars are different. Every dollar, they declare, has something logically independent of its functionalistic exchange powers, which we might call its *vim*. So defined, the *vim* of each dollar is guaranteed to elude the theories of economists forever, but we have no reason to believe in it—aside from the heartfelt hunches of those naive Americans, which can be explained without being honored.

Some participants in the consciousness debates simply demand, flat out, that their intuitions about phenomenal properties are a nonnegotiable starting point for any science of consciousness. Such a conviction must be considered an interesting symptom, deserving a diagnosis, a datum that any science of consciousness must account for, in the same spirit that economists and psychologists might set out to explain why it is that so many people succumb to the potent illusion that money has intrinsic value.

There are many properties of conscious states that can and should be subjected to further scientific investigation right now, and once we get accounts of them in place, we may well find that they satisfy us as an explanation of what consciousness is. After all, this is what has happened in the case of the erstwhile mystery of what *life* is. Vitalism—the insistence that there is some big, mysterious extra ingredient in all living things—turns out to have been not a deep insight but a failure of imagination. Inspired by that happy success story, we can proceed with our scientific exploration of consciousness. If the day arrives when all these acknowledged debts are paid and we plainly see that something big is missing (it should stick out like a sore thumb at some point, if it is really important) those with the unshakable hunch will get to say they told us so. In the meantime, they can worry about how to fend off the diagnosis that they, like the vitalists before them, have been misled by an illusion.

References

Akins, Kathleen. 1993. "What Is It Like to Be Boring and Myopic?" In E. Dahlbom, ed., *Dennett and His Critics: Demystifying Mind*, pp. 124–160. Oxford: Blackwell.

———. 2001. "More Than Mere Coloring: A Dialogue between Philosophy and Neuroscience on the Nature of Spectral Vision." In S. Fitzpatrick and J. T. Breuer, eds., *Carving our Destiny*. Washington, D.C.: Joseph Henry Press.

Alter, Torin, and Sven Walter, eds. 2005. *Phenomenal Concepts and Phenomenal Knowledge: New Essays on Consciousness and Physicalism*. New York: Oxford University Press.

Asch, Solomon. 1958. "Effects of Group Pressure upon the Modification and Distortion of Judgements." In E. E. Maccoby, T. M. Newcomb, and E. L. Hartely, eds., *Readings in Social Psychology*, pp. 174–181. New York: Henry Holt.

Baars, B. 1988. *A Cognitive Theory of Consciousness*. Cambridge: Cambridge University Press.

Block, N. 2001. "Paradox and Cross Purposes in Recent Work on Consciousness." In Dehaene and Naccache 2001, pp. 197–219.

Bower, G., and Clapper, J. 1989. "Experimental Methods in Cognitive Science." In Posner 1989, pp. 245–300.

Bringsjord, Selmer. 1994. "Dennett versus Searle on Cognitive Science: It All Comes Down to Zombies and Searle Is Right." Paper presented at the APA, December 1994. This paper has been developed into "The Zombie Attack on the Computational Conception of Mind," *Philosophy and Phenomenological Research* 59 (1) (1999): 41–69. Available online at http://www.rpi.edu/~brings/SELPAP/ZOMBIES/zomb.htm.

Brook, Andrew. 2000. "Judgments and Drafts Eight Years Later." In D. Ross and A. Brook, eds., *Dennett's Philosophy: A Comprehensive Assessment.* Cambridge, Mass.: The MIT Press.

Burgess, P. W., Baxter, D., Rose, M., and Alderman, N. 1996. "Delusional Paramnesic Misidentification." In P. W. Halligan and J. C. Marshall, eds., *Method in Madness: Case Studies in Cognitive Neuropsychiatry*, pp. 51–78. Hove: Psychology Press.

Chalmers, David. 1995. "Facing Up to the Problem of Consciousness." *Journal of Consciousness Studies* 2: 200–219.

———. 1996. *The Conscious Mind.* New York: Oxford University Press.

———. 1999. "First-Person Methods in the Science of Consciousness." *Consciousness Bulletin* (fall): 8–11.

———. Reply to Searle. Available on his Web site, http://www.u.arizona.edu/~chalmers/discussions.html.

Chomsky, Noam. 1994. "Naturalism and Dualism in the Study of Mind and Language." *International Journal of Philosophical Studes* 2: 181–209.

Crick, Francis. 1994. *The Astonishing Hypothesis: The Scientific Search for the Soul.* New York: Scribner.

Damasio, Antonio. 1999. *A Feeling for What Happens.* New York: Harcourt Brace.

Davidson, D. 1987. "Knowing One's Own Mind." *Proceedings and Addresses of the American Philosophical Association* 60: 441–458.

Debner, J. A., and L. L. Jacoby. 1994. "Unconscious Perception: Attention, Awareness, and Controls." *Journal of Experimental Psychology: Learning Memory and Cognition* 20: 304–317.

Dehaene, S., and L. Naccache, eds. 2001. Special issue of *Cognition*, vol. 79, *The Cognitive Neuroscience of Consciousness*. Reprinted in 2001 by The MIT Press.

Dennett, Daniel. 1971. "Intentional Systems." *Journal of Philosophy* 68: 87–106.

————. 1978. *Brainstorms: Philosophical Essays on Mind and Psychology.* Cambridge, Mass.: The MIT Press/A Bradford Book.

————. 1979. "On the Absence of Phenomenology." In D. Gustafson and B. Tapscott, eds., *Body, Mind, and Method: Essays in Honor of Virgil C. Aldrich*. Dordrecht: D. Reidel.

————. 1982. "How to Study Consciousness Empirically: Or, Nothing Comes to Mind." *Synthese* 53: 159–180.

————. 1987. *The Intentional Stance*. Cambridge, Mass.: The MIT Press/A Bradford Book.

————. 1988. "Quining Qualia." In Marcel and Bisiach, eds., *Consciousness in Contemporary Science*. Cambridge: Cambridge University Press.

————. 1991. *Consciousness Explained*. Boston and New York: Little, Brown.

————. 1994a. "Get Real." Reply to my critics in *Philosophical Topics* 22: 505–568.

————. 1994b. "Real Consciousness." In A. Revonsuo and M. Kamppinen, eds., *Consciousness in Philosophy and Cognitive Neuroscience*. Hillsdale, N.J.: Lawrence Erlbaum.

————. 1994c. "Instead of Qualia." In A. Revonsuo and M. Kamppinen, eds., *Consciousness in Philosophy and Cognitive Neuroscience*. Hillsdale, N.J.: Lawrence Erlbaum.

————. 1995a. "Overworking the Hippocampus," commentary on Jeffrey Gray. *Behavioral and Brain Sciences* 18: 677–678.

————. 1995b. "The Unimagined Preposterousness of Zombies." *Journal of Consciousness Studies* 2: 322–336.

———. 1996a. "Cow-Sharks, Magnets, and Swampman." *Mind and Language* 11 (1): 76–77.

———. 1996b. "Consciousness: More Like Fame Than Television" in German translation: "Bewusstsein hat mehr mit Ruhm als mit Fernsehen zu tun." In Christa Maar, Ernst Pöppel, and Thomas Christaller, eds., *Die Technik auf dem Weg zur Seele*. Berlin: Rowohlt.

———. 1996c. "Facing Backwards on the Problem of Consciousness," commentary on Chalmers for *Journal of Consciousness Studies* 3 (1, special issue, part 2): 4–6. Reprinted in *Explaining Consciousness: The "Hard Problem,"* ed. Jonathan Shear. Cambridge, Mass.: The MIT Press/A Bradford Book, 1997.

———. 1998a. "The Myth of Double Transduction." In S. Hameroff, A. W. Kaszniak, and A. C. Scott, eds., *Toward a Science of Consciousness II: The Second Tucson Discussions and Debates*, pp. 97–107. Cambridge, Mass.: The MIT Press.

———. 1998b."No Bridge Over the Stream of Consciousness," commentary on Pessoa et al. *Behavioral and Brain Sciences* 21: 753–754.

———. 1999. "Intrinsic Changes in Experience: Swift and Enormous," commentary on Palmer. *Behavioral and Brain Sciences* 22 (6): 951.

———. 2000a. "The Case for Rorts." In R. B. Brandom, ed., *Rorty and His Critics*, pp. 89–108. Oxford: Blackwell.

———. 2000b. "It's Not a Bug, It's a Feature," commentary on Humphrey. *Journal of Consciousness Studies* 7: 25–27.

———. 2001a. "Are We Explaining Consciousness Yet?" *Cognition* 79: 221–237.

———. 2001b. "The Zombic Hunch: Extinction of an Intuition?" In Anthony O'Hear, ed., *Philosophy at the New Millennium* (*Royal Institute of Philosophy Supplement* 48: 27–43). Cambridge: Cambridge University Press.

———. 2001c. "Explaining the 'Magic' of Consciousness." In *Exploring Consciousness, Humanities, Natural Science, Religion*, Proceedings of the

International Symposium, Milano, November 19–20, 2001 (published in December 2002, Fondazione Carlo Erba), pp. 47–58). Reprinted in J. Laszlo, T. Bereczkei, and C. Pleh, eds., *Journal of Cultural and Evolutionary Psychology* 1(2003): 7–19.

———. 2001d. "Surprise, Surprise," commentary on O'Regan and Noë. *Behavioral and Brain Sciences* 24 (5): 982.

———. 2002a. "How Could I Be Wrong? How Wrong Could I Be?" *Journal of Consciousness Studies* 9 (5–6), special issue: "Is The Visual World a Grand Illusion?" Alva Noë, ed., pp. 13–16.

———. 2002b. "Does Your Brain Use the Images in It, and If So, How?", commentary on Pylyshyn. *Behavioral and Brain Sciences* 25 (2): 189–190.

———. 2003a. *Freedom Evolves*. New York: Viking Penguin.

———. 2003b. "Who's on First? Heterophenomenology Explained." *Journal of Consciousness Studies*, special issue: "Trusting the Subject? (Part 1)," 10 (nos. 9–10, October): 19–30. Also appears in A. Jack and A Roepstorff, eds., *Trusting the Subject?* vol. 1, pp. 19–30. Imprint Academic, 2003.

———. 2003c. "Look Out for the Dirty Baby," peer commentary on Baars. *Journal of Consciousness Studies*, "The Double Life of B. F. Skinner," 10 (1): 31–33.

———. 2003d. "Making Ourselves at Home in Our Machines," review of Wegner, *The Illusion of Conscious Will*, The MIT Press, 2002. *Journal of Mathematical Psychology* 47: 101–104.

Dennett, Daniel, and M. Kinsbourne. 1992. "Time and the Observer: The Where and When of Consciousness in the Brain." *Behavioral and Brain Sciences* 15: 183–247.

DeSimone, R., and J. Duncan. 1995. "Neural Mechanisms of Selective Visual Attention." *Annual Review of Neuroscience* 18:193–222.

Driver, J., and P. Vuilleumer. 2001. "Perceptual Awareness and Its Loss in Unilateral Neglect and Extinction." In Dehaene and Naccache 2001, pp. 39–88.

Ellis, H. D., Lewis, M. B. 2001. "Capgras Delusion: A Window on Face Recognition." *Trends in Cognitive Science* 5: 149–156.

Fodor, Jerry. 1998. "Review of Steven Pinker's *How the Mind Works*, and Henry Plotkin's *Evolution in Mind*." *London Review of Books*, January 22. Reprinted in J. Fodor, *In Critical Condition* (Cambridge, Mass.: The MIT Press/A Bradford Book, 1998).

Frith, C. D. 2000. "The Role of Dorsolateral Prefontal Cortex in the Selection of Action, as Revealed by Functional Imaging." In S. Monsell and J. Driver (eds.), *Control of Cognitive Processes: Attention and Performance*, vol. 18. Cambridge, Mass.: The MIT Press.

Goldman, Alvin. 1997. "Science, Publicity, and Consciousness." *Philosophy of Science* 64 (4): 525–545.

———. 2000. "Can Science Know When You're Conscious?" *Journal of Consciousness Studies* 7 (5): 2–22.

Graham, George, and Terence Horgan. 2000. "Mary Mary Quite Contrary." *Philosophical Studies* 99: 59–74.

Gray, Jeffrey. 1995. "The Contents of Consciousness: A Neuropsychological Conjecture." *Behavioral and Brain Sciences* 18 (4): 659–722.

Hayes, Patrick. 1978. "The Naive Physics Manifesto." In D. Michie, ed., *Expert Systems in the Microelectronic Age*. Edinburgh: Edinburgh University Press.

Hilliard, John Northern. 1938. *Card Magic*. Minneapolis: Carl W. Jones.

Hofstadter, Douglas. 1979. *Gödel, Escher, Bach: An Eternal Golden Braid*. New York: Basic Books.

———. 1981. "Reflections." In Hofstadter and Dennett, ed., *The Mind's I*, p. 375. New York: Basic Books.

Hofstadter, Douglas, and Daniel Dennett, eds. 1981. *The Mind's I*, New York: Basic Books.

Humphrey, Nicholas. 2000. "How to Solve the Mind–Body Problem" (with commentaries and a reply by the author). *Journal of Consciousness*

Studies 7: 5–20. (Also available as a book, *How to Solve the Mind–Body Problem*.)

Hurley, Susan. 1998. *Consciousness in Action*. Cambridge, Mass.: Harvard University Press.

Jack, Anthony I., and T. Shallice. 2001. "Introspective Physicalism as an Approach to the Science of Consciousness." In Dehaene and Naccache 2001, pp. 135–159.

Jackson, Frank. 1982. "Epiphenomenal Qualia." *Philosophical Quarterly* 32: 27–36.

Janet, Pierre. 1942. *Les Dissolutions de la Memoire*. Quoted in Tolland, *Disorders of Memory*, 1968, p. 152.

Kanwisher, N. 2001. "Neural Events and Perceptual Awareness." In Dehaene and Naccache 2001, pp. 89–113.

Kawamura, Y., and Kare, M. R., eds. 1987. *Umami: A Basic Taste*. New York: Dekker.

Levine, Joseph. 1983. "Materialism and Qualia: The Explanatory Gap." *Pacific Philosophical Quarterly* 64: 354–361.

————. 1994. "Out of the Closet: A Qualophile Confronts Qualophobia." *Philosophical Topics* 22: 107–126.

Lycan, William. 1987. *Consciousness*. Cambridge, Mass.: The MIT Press.

————. 1996. *Consciousness and Experience*. Cambridge, Mass.: The MIT Press.

————. Forthcoming. "Perspectival Representation and the Knowledge Argument." In Q. Smith and A. Jokic (eds.), *Consciousness: New Philosophical Essays*. Oxford: Oxford University Press.

McCarthy, John. 1959. Discussion of Oliver Selfridge, "Pandemonium: A Paradigm for Learning." In *Symposium on the Mechanization of Thought Processes*. London: H. M. Stationery Office.

McConnell, Jeff. 1994. "In Defense of the Knowledge Argument." *Philosophical Topics* 22: 157–197.

McGeer, Victoria. 2003. "The Trouble with Mary." *Pacific Philosophical Quarterly* 84 (4): 384–393.

McGinn, Colin. 1999. *The Mysterious Flame: Conscious Minds in a Material World.* New York: Basic Books.

Merikle, Philip M., Daniel Smilek, and John D. Eastwood. 2001. "Perception without Awareness: Perspectives from Cognitive Psychology." *Cognition* 79: 115–134.

Nagel, Thomas. 1974. "What Is It Like to Be a Bat?" *Philosophical Review* 83: 435–450.

———. 1979. *Mortal Questions.* Cambridge: Cambridge University Press.

———. 1998. "Conceiving the Impossible and the Mind–Body Problem." *Philosophy* 73: 337–352.

O'Craven, K. M., B. R. Rosen, K. K. Kwong, A. Treisman, and R. L. Savoy. 1997. "Voluntary Attention Modulates fMRI Activity in Human MT/MST." *Neuron* 18: 591–598.

Palmer, S. 1999. "Color, Consciousness, and the Isomorphism Constraint." *Behavioral and Brain Sciences* 22: 923–989.

Panskepp, J. 1998. *Affective Neuroscience: The Foundations of Human and Animal Emotions.* Oxford and New York: Oxford University Press.

Parvizi, Josef, and Antonio Damasio. 2001. "Consciousness and the Brain Stem." *Cognition* 79: 135–159.

Penrose, Roger. 1989. *The Emperor's New Mind: Concerning Computers, Minds, and the Laws of Physics.* Oxford: Oxford University Press.

Pessoa, L., E. Thompson, and A. Noë. 1998. "Finding Out about Filling In: A Guide to Perceptual Completion for Visual Science and the Philosophy of Perception." *Behavioral and Brain Sciences* 21: 723–748.

Petitot, J., F. Varela, B. Pachoud, and J.-M. Roy. 1999. *Naturalizing Phenomenology: Issues in Contemporary Phenomenology and Cognitive Science.* Stanford: Stanford University Press.

Posner, M. I., ed. 1989. *Foundations of Cognitive Science*. Cambridge, Mass.: The MIT Press.

Powers, Richard. 1995. *Galatea 2.2*. New York: Harpers.

Pylyshyn, Z. 2002. "Mental Imagery: In Search of a Theory." *Behavioral and Brain Sciences* 25: 157–237.

Rensink, R. A., J. K. O'Regan, and J. J. Clark. 1997. "To See or Not to See: The Need for Attention to Perceive Changes in Scenes." *Psychological Science* 8 (5): 368–373.

Robinson, Howard. 1993. "Dennett on the Knowledge Argument." *Analysis* 53: 174–177.

Rolls, E. T., and T. Yamamoto, eds. 2001. *Sensory Neuron* 3: "Glutamate Receptors: Taste Perception and the Brain."

Rozin, Paul. 1976. "The Evolution of Intelligence and Access to the Cognitive Unconscious." *Progress in Psychobiology and Physiological Psychology*, vol. 6, pp. 245–280. New York: Academic Press.

Searle, John. 1980. "Minds, Brains, and Programs." *Behavioral and Brain Sciences* 3: 417–458.

———. 1992. *The Rediscovery of the Mind*. Cambridge, Mass.: The MIT Press.

Shepard, Roger, and J. Metzler. 1971. "Mental Rotation of Three-Dimensional Objects." *Science* 171: 701–703.

Siegel, Lee. 1991. *Net of Magic: Wonders and Deceptions in India*. Chicago: University of Chicago Press.

Smiley, Jane. 2004. "Excerpts from Jane Smiley's New Book." *Practical Horseman* (June): 44–68.

Smith, S. D., and P. M. Merikle. 1999. "Assessing the Duration of Memory for Information Perceived without Awareness." Poster presented at the Third Annual Meeting of the Association for the Scientific Study of Consciousness, London, Ontario, Canada, June 1999.

Strawson, Galen. 1999. "Little Gray Cells." *New York Times Book Review*, July 7, 1999, p. 13.

Thompson, Evan. 2001. "Empathy and Consciousness." *Journal of Consciousness Studies* 8: 1–33.

Tye, Michael. 1995. *Ten Problems of Consciousness*. Cambridge, Mass.: The MIT Press.

Varela, Francisco, and Jonathan Shear. 1999. "First-Person Methodologies: What, Why, How?" *Journal of Consciousness Studies* 6 (203): 1–14.

Voorhees, Burton. 2000. "Dennett and the Deep Blue Sea." *Journal of Consciousness Studies* 7: 53–69.

Weiskrantz, L. 1998. "Consciousness and Commentaries." In S. R. Hameroff, A. W. Kaszniak, and A. C. Scott (eds.), *Toward a Science of Consciousness II: The Second Tucson Discussions and Debates*, pp. 11–25. Cambridge, Mass.: The MIT Press.

Wright, Robert. 2000. *Nonzero: The Logic of Human Destiny*. New York: Pantheon.

Zhao, Grace Q., Yifeng Zhang, Mark A. Hoon, Jayaram Chandrashekar, Isolde Erlenbach, Nicholas J. P. Ryba, and Charles S. Zuker. 2003. "The Receptors for Mammalian Sweet and Umami Taste." *Cell* 115: 255–266.

Index